Coring Methods and Systems

Rahman Ashena · Gerhard Thonhauser

Coring Methods and Systems

 Springer

Rahman Ashena
Department of Petroleum Engineering
Montanuniversität Leoben
Leoben
Austria

Gerhard Thonhauser
Department of Petroleum Engineering
Montanuniversität Leoben
Leoben
Austria

ISBN 978-3-030-08524-7   ISBN 978-3-319-77733-7   (eBook)
https://doi.org/10.1007/978-3-319-77733-7

Printed on acid-free paper

This Springer imprint is published by the registered company Springer International Publishing AG
part of Springer Nature
The registered company address is: Gewerbestrasse 11, 6330 Cham, Switzerland

Acknowledgements

As *Coring Methods and Systems* has greatly benefited from technical consultation contributions of several individuals, we would like to express special words of appreciation to *Mr. George Williamson* and *Mr. Audun Kvinnesland*, from *Baker Hughes GE*; *Mr. Ian Ross* (for his continuous sincere support) and also *Mr. Corin Lewsey* from *National Oilwell Varco; Mr. Rudolf Nikolaus Knezevic* from *OMV*; *Mr. Jonathan Rylance* and *Mr. Alex Schellenberg* from *Reservoir Group*; *Mr. Mike Burns* from *Halliburton*; *Mr. Chris Daws* and *Mr. George Tophinke* from *Sandvik, Australia*; *Mr. Jos Nutbroek* from *Boart Longyear*; *Ms. Izaskun Zubizarreta* from *Lloyds Register*; *Mr. Tony Kennaird* from *Core Laboratories Australia; Mr. Yuichi Shinmoto from JAMSTEC; Prof. Walter Vortisch* and *Prof. Michael Prohaska* from *Montanuniversität Leoben*; and *Dr. Christian Koller* from *TDE Group*. We kindly appreciate *Mr. Ian Ross* for his help on preparation of some of the schematics. Special thanks go to the companies which permitted us to use some of their figures and tool specifications: *Baker Hughes GE, National Oilwell Varco (NOV), Reservoir Group (formerly ALS Corpro), Halliburton, Sandvik*, and *Boart Longyear*.

Finally, we appreciate *Prof. Reza Rezaee* from *Curtin University* for reviewing the book and providing constructive comments.

Rahman Ashena
Gerhard Thonhauser

Contents

List of Figures

List of Tables

About the Authors

Dr. Rahman Ashena (Soroush) completed his Master and Ph.D. degrees in Petroleum Well Engineering respectively at Curtin University (Australia) in 2009, and Montanuniversität Leoben (Austria) in 2017. The topic of his Ph.D. dissertation was on an innovative geomechanics approach in core tripping. Backed up with around five years of field work experience as a senior well completion/workover and drilling supervisor engineer, Ashena is specialized in well operations and engineering (including well control, geomechanics, wellbore fluid flow modeling, well integrity management, and coring). In addition to his industry experience, Ashena has more than seven years' experience of lecturing drilling and completion courses. Ashena has published several technical papers with international publishers including SPE, Elsevier, EAGE, and Taylor & Francis. He has also authored a book chapter in "Artificial Intelligent Approaches in Petroleum Geosciences", Springer and a handbook on cement bond logging and evaluation.

Prof. Gerhard Thonhauser has 20 years of experience providing the petroleum industry with the evaluation, analysis, and management of drilling and well-related data. He is the founder of TDE Group GmbH servicing the international industry with drilling performance analysis and benchmarking, as well as software, electronics, and equipment development and manufacturing. Since 2003, he is Professor at the Montanuniversität Leoben, Austria and holds the Chair of Drilling and Completion Engineering and acts as the Head of the Petroleum Engineering Department. His areas of research and development include "Drilling Process Monitoring and Analysis" combining modeling, advanced monitoring and sensing, and drilling analysis to optimize and automate the drilling process and to improve learning and knowledge management in drilling organizations. He has authored and co-authored around 100 publications in the area of drilling and completion engineering and data science. Since 2009, he is the head of "Environmental Friendly Drilling Systems" initiative in Europe focusing on drilling and completion technologies with "Zero Harmful Environmental Impact". He holds a Masters and a Ph.D. degree in Petroleum Engineering from the Montanuniversität Leoben, Austria. He is also an active member of SPE.

Chapter 1
Introduction

1.1 General Introduction

Coring is the method of providing rock samples from subsurface formations. The analysis of the samples can potentially provide reliable information about the rock and fluid properties. It provides reservoir properties in much smaller scale than the other geoscience or exploration techniques (such as well logging or seismic surveys). However, it is considered more reliable and vital in exploration (either in the petroleum, mining or any geo-related industry) as the data obtained from the core analysis are considered as the ground truth. Thus, core data is used for the purpose of calibrating the data from other exploration methods. In the petroleum industry, for example, the main goal of coring is to practically identify formations with a commercial scale of oil and gas content.

A successful coring job is the one which achieves the objectives of the job. Considering the objectives, the criteria required for coring and the obtained cores determine the right coring methods or systems. If these methods are correctly selected and implemented, they can supply undisputed results about the formation rock and fluid properties. Undoubtedly, coring had originally faced a lot of challenges including cost, technical problems causing core damage due to coring causing invalidity of cores, cores with inadequate geological data, etc. Coring in unconventional reservoirs with typically unconsolidated formation types faced additional challenges in the past. This has required more innovative methods to be developed and applied. Therefore, coring methods and systems with core barrels of various types and components have been developed for use based on the formation types and coring goals. Recently, lessons learned from other industries particularly mining have also greatly contributed to the development of the wireline continuous coring as a lower cost rival of conventional coring. This system is still being enhanced to remove its drawbacks and obtain samples with greater size and quality.

Core damage prevention and mitigation during coring is also an important concentration in the industry to make coring a more reliable exploration technique.

© Springer International Publishing AG, part of Springer Nature 2018
R. Ashena and G. Thonhauser, *Coring Methods and Systems*,
https://doi.org/10.1007/978-3-319-77733-7_1

The industry has addressed this topic by applying invasion–mitigation methods (such as low-invasion system, gel coring, and sponge coring) and core-drilling optimization. Special attention to core tripping or retrieval has been recently paid in order to mitigate the induced gas expansion damage which creates microfractures, which includes some authors' findings.

In pursuit of other methods and systems, for the purpose of increasing the amount of measurable geological data from the core and the amount of measured data while coring, respectively, oriented coring and Logging-While-Coring (LWC) systems have been developed. Oriented coring is currently considered invaluable by the industry for providing significant geological understanding of the formations. LWC has shown to be potentially capable of revolutionizing the coring operations particularly in terms of enhanced monitoring, control, and decision-making, provided that the mechanical issues corresponding to the size of the electronics and the telemetry issues can be resolved. Next, for pressure/in situ coring with lower level of safety issues, in recent years some newly developed systems have been developed, which faced an initial warm welcome by the market. The industry is currently showing growing attention to motor coring for controlling the coring operations in directional wellbores as nowadays most wells may have a deviated wellbore. Coiled-tubing coring is also considered as a potentially revolutionizing method with reduced costs. The industry is currently considering mini-coring to obtain greater information about the cuttings than what is obtainable by the cuttings. Depending on the rock sampling requirements and conditions, the industry has also practiced combining several coring systems with the fundamental methods.

The discussion about the state-of-the-art coring methods and systems including their core barrel types and components was really missing in the literature (particularly for the petroleum industry) and there comes the incentive of the authors for this book.

1.2 Objectives

In this coring book, the following objectives are followed:

1. Explaining the reasons for utilizing coring in comparison of other exploration techniques, coring uncertainties and challenges, and explanation of a justified coring job;
2. Description of fundamental coring methods (conventional, wireline continuous, and sidewall);
3. Description of core barrel components, types, and designs in conventional and wireline continuous coring methods (in petroleum and mining industries) and their comparison;
4. Explanation and description of the state-of-the-art coring methods and systems;

5. Core damage investigation and mitigation due to coring fluid invasion, coring process, tripping, etc.;
6. Recommended handling procedures;
7. Introduction of some coring providers, detailed specifications of their products, and patents.

1.3   Scope

This book consists of 15 chapters to cover the state-of-the-art methods and systems of coring (including the core barrel components) and handling. It is noted that core preservation and laboratory analysis are out of the scope of this book. The scope is explained as follows: First, this chapter provides an introduction to coring, the objectives, and an outline of the scope.

In Chap. 2, following a definition given for coring, a justification for coring is given having considered the inherent economic and technical challenges. To make this tangible, a typical model of justified coring is presented which can lead to reliable coring with mitigated costs.

In Chap. 3, a brief description of fundamental coring methods (*conventional, wireline continuous*, and *sidewall*) is presented. It is noted that as this handbook is purposefully aiming at main exploration investigation of prospective reservoirs, detailed discussion about sidewall coring is out of the scope. This is because this method is not comparatively as efficient exploration as the others to identify the formation properties due to the limitation of its obtainable data.

In Chap. 4, various types of core barrels and systems commonly used in the petroleum industry are introduced and schematically depicted (*conventional, wireline continuous, slim hole, high torque, jam-indicating, Jam-mitigation (i.e., antijamming* and *full-closure), invasion–mitigation, oriented, pressure/in situ, Logging-While-Coring*, and *motor coring*). Next, both for the conventional and wireline continuous methods, the components and functions of their outer tube/barrel and inner tube/barrel assemblies utilized in the petroleum industry are depicted and illustrated. In the following chapter, the detailed components and schematics of the core barrels utilized in the mining industry are depicted schematically, which have some slight differences from the petroleum wireline continuous coring systems.

In Chap. 5, the conventional coring method is explained including its operations procedure, operations, and a detailed schematic showing the working mechanism. Finally, the challenges corresponding to this method are investigated. This chapter shows how coring has been originally implemented.

In Chap. 6, wireline continuous coring is introduced as the other fundamental method discussed in this book. This method, which is the heritage of the mining industry, is increasingly growing in the petroleum industry. In addition, the operational procedures of the method for both its modes (coring and drilling) and also the latching mechanism are covered in addition to the schematic illustration of the method. It is noted that wireline continuous method is becoming popular because of its main feature/possibility of switching from one mode to the other without requiring any conventional tripping. Next, the navigated version of wireline coring is discussed as an almost dated method which used to measure the gamma ray during coring.

As core damage mitigation is one of the main coring objectives, it is discussed in Chap. 7 for the mud invasion and in Chap. 8 for the mechanical damage. Therefore, in Chap. 7, invasion–mitigation coring systems (*low-invasion*, *gel coring*, and *sponge coring*) are discussed as a significant category of coring systems. This chapter investigates the methods of mitigating the invasion-driven core damage and core quality enhancement during its cutting and entry into the inner tube. A low-invasion coring system requires using a low-invasion inner tube lower shoe, core bit, coring mud, and also an optimized core-drilling. It is noted that the use of a non-split inner tube liner inside the inner tube (called *triple-tube barrel*) or a split liner inside the steel inner tube are options to mitigate the mud invasion. Next, the gel coring system is introduced which contains pistons and gel barrel/tube to encapsulate the core as it enters the inner tube. Finally, sponge coring system is discussed which is used to absorb the liquid hydrocarbons pushed out of the core by the gas expansion during tripping.

In Chap. 8, the sources of mechanical core damage are investigated and attributed to its cutting and also its tripping and retrieval to the surface. To prevent or mitigate, geomechanical modeling and simulation are required. Therefore, this chapter depicts the recently developed methods and models.

In Chap. 9, first, the monitoring of the operational drilling parameters and their roles in detection of incidents such as jamming are investigated for both the conventional and wireline methods. Next, the jamming mitigation methods/systems, which consist of antijamming and full-closure systems, are discussed as important features that can be combined with a fundamental coring method (either a conventional or wireline continuous method). Both systems are suitable for coring in weak unconsolidated and also unconventional reservoirs. The antijamming system applies several concentric sleeves which enables continued coring in jam-prone formations without unprecedented termination of the job. The full-closure method constitutes (1) utilizing a slick entry inner tube, and (2) activating an additional ball to be dropped on its corresponding seat which raises the inner tube and activates the full-closure shoe downhole.

In Chap. 10, oriented coring is discussed as a contributing feature added to mostly conventional and even wireline coring methods. This incorporates a survey tool for gravitational and magnetic measurements, and also the *scribe knives* at the inner tube lower shoe. Therefore, enhanced core geological properties can be measured including the strata dip angle and strike, fracture orientation, etc.

In Chap. 11, the pressure/in situ coring system is discussed as a recently applied successful innovation in rock sampling. Using this method, the core sample is retrieved under full or part of the in situ pressure. This, in addition, enables the pore fluids (particularly gas) which are being expelled and expanded from the sample during tripping, to be captured. This is ideal for gas volume measurements and also reservoir engineering tests. However, some considerable safety concerns regarding handling high-pressure barrels at the surface still remain.

In Chap. 12, Logging-While-Coring (LWC) is brought up as another recently field-tested innovative feature in coring. This system utilizes downhole sensors to measure in situ properties of the sample. The geometry of the sensors must be modified to be placed in the annulus between inner and outer tubes to measure formation properties just at the bit. This feature provides a more accurate decision-making capability during the cutting and tripping phases. LWC can be added as a feature to fundamental coring methods (e.g., the wireline continuous method). The data measured by this feature can be recorded in the battery to be retrieved when the inner tube is recovered, or it can be sent to the surface in a real-time manner.

In Chap. 13, some other coring systems are discussed which are less common. These comprise motor coring, underbalanced coring, and coiled-tubing coring among which motor coring is the most in practice.

In Chap. 14, detailed core handling procedures are discussed with emphasis on the proper practices in order to prevent or mitigate the core damage during transportation to the rig site lab or central lab.

Finally, in Chap. 15, the list of some coring providers in the petroleum and mining sectors along with the detailed specifications of their products and also some important patents regarding the innovations in the systems have been provided.

Chapter 2
Justified Coring

In this chapter, first it is necessary to provide a definition to coring as an exploration method. Next, the uncertainties and challenges of this method are investigated including its validity and cost. Then, coring is compared with other exploration methods. Finally, its justification is investigated from a practical point of view.

2.1 Definition

Generally, the objective of drilling an oil or gas well is to locate reservoir formations with a commercial accumulation of hydrocarbon. During the course of drilling, therefore, additional precise information may be necessary concerning the lithology and fluid type of the formation. This enables in-time decision-making to complete the well before spending additional expenses required for subsequent well completion and production. In order to obtain an idea of these formation rock and fluid properties, several exploration methods are available (including coring, well logging, seismic survey, etc.)

Coring is the extraction of cylindrical rock samples from their native state for the purpose of their physical examination at the surface. Cores are large compared to normal cuttings sizes, which is an advantage for the examination. Coring could also be defined as cutting an annulus or washer shape around a central cylinder of rock. In terms of objectives, it is performed in order to provide qualitative and quantitative geological and mechanical data required for reservoir characterization and decision-making. In addition to the reservoir engineering objectives, core samples with sufficient diameter and length provide invaluable data about geological

© Springer International Publishing AG, part of Springer Nature 2018
R. Ashena and G. Thonhauser, *Coring Methods and Systems*,
https://doi.org/10.1007/978-3-319-77733-7_2

bedding, formation dip and strike, stratigraphy, mineralogy, fractures, etc. Coring is accomplished by virtue of two concentric pipes, i.e., the inner tube and outer tube. The inner tube holds the core to protect it from damage during coring, core retrieval, and handling processes. There could be an additional liner inside the inner tube for better protection and containerization of the core rock material. The outer tube assembly (together with the stabilizers and the overlying drill collars and jar) act as the Bottomhole Assembly (BHA) and conduit to connect the drill pipes to the core bit/head.

Following the coring operations, the cores are retrieved and handled at the surface to be sent either to the rig site laboratory for some basic experiments or to the central laboratory. From a reservoir engineering point of view, the most valuable data obtained from the core analysis are the basic routine measurements of porosity, permeability, oil, gas, and water saturations. Following the rig site lab tests (if it is an option), core preservation is performed prior to its dispatch to the central laboratory for further sophisticated core analysis.

2.2 Uncertainties and Challenges

The information obtained by the analysis of the retrieved core samples will be used by a broad variety of petroleum experts (well completion, reservoir engineers, geologists, and petrophysicists) to evaluate the rock and fluid properties of the zone of interest. This enables the decision-making about the economic potential of the reservoir/well. If it is an exploration well, the decision-making is corresponding to proceeding to either development drilling in the field or moving away in pursuit of hydrocarbon in another region. If the prospective well is developmental, the decision-making would revolve around completing or abandoning the well. If well logging could also be performed while coring, the logging data would contribute to coring optimizing and better controlling the coring operations. In practice, coring operations are applied more for exploration drilling rather than development wells.

Based on the aforementioned, coring as a method of exploration has continued to be an accepted exploration method as it contributes to finding a lot of information (listed in Table 2.1). Like any other exploration method, it has some advantages and disadvantages (Table 2.2). There are some disadvantages or challenges (the validity of the analysis results and the cost) which may make the management doubtful about the value of coring as an optimal exploration technique, particularly in hard economic situations (e.g., of low oil prices). The coring challenges are discussed in the next section.

Table 2.1 Information and data obtained using core analysis

Category	Information	Figure
Well completion	Decision if we can continue completing the well Decision if acidizing, hydraulic-fracturing, etc. is essential. If so, their designs	
Reservoir engineering (RCAL[a])	Porosity	
	Permeability	
	Porosity–permeability relationship and *Hydraulic flow unit* discrimination	b
	Fluid saturation	
	Grain density	
	Oil–water contact/gas–water contact	
	Lithological description	
Reservoir engineering (SCAL[c])	Relative permeability	
	Capillary pressure	
	Electrical properties	
Geological studies	Lithology	
	Rock type	
	Depositional environment	
	Pore type	
	Mineralogy (and shale CEC)/geochemistry	
	Diagenesis	
	Fluorescence	d
	Formation age and sequence stratigraphy	
	Fracture studies	
Petrophysical analysis (after calibration by cores)	Calibrating log data	
	GR response	
	Resistivity	
	Density	
	Sonic velocity	e

[a]Routine Core Analysis
[b]https://www.onepetro.org/conference-paper/SPE-180239-MS?sort=&start=0&q=180239&from_year=&peer_reviewed=&published_between=&fromSearchResults=true&to_year=&rows=10#
[c]Special Core Analysis
[d]http://www.pdgm.com/getmedia/f6caf52c-ce8b-4144-b787-9e4385d71166/geomage.jpg.aspx?width=1024&height=650&ext=.jpg (accessed on September 20, 2015)
[e]http://www.oman.petrotel.com/?page=petrophysical (accessed on September 20, 2015)

Table 2.2 General advantages and disadvantages of coring as an exploration technique

	Coring	
	Advantages	Disadvantages
1.	Provides Routine Core Analysis (RCAL) data (porosity, permeability, water and hydrocarbon saturation)	Validity issues such as change of rock and fluid properties from the in situ to the atmosphere, hydrocarbon fluid escape and mud invasion, and stress release and hysteresis
2.	Provides Special Core Analysis (SCAL) data (relative permeability, capillary pressure, wettability, etc.)	
3.	Fluid volume in place	
4.	Provides geomechanical properties (strength, Poisson ratio, etc.)	High cost

2.2.1 Validity

There are some arguments against coring in regard to validity of the core data obtained compared with other alternatives. Thus, there are some uncertainties about coring validity including the following:

- The fluid and rock properties because of the change from the in situ to the atmosphere, causing hydrocarbon fluid escape, and stress release and hysteresis, etc.
- If the quality, quantity, and conditions of the recovered cores have been retained (as the damage occurring to the core during its drilling, retrieval, and handling can lead to the deviation of the core analysis results from the real in situ conditions). In other words, there is an uncertainty if the properties of the samples retrieved are really representative of the reservoir.
- If the results of core analysis could be extrapolated to the entire reservoir, particularly in heterogeneous reservoirs.
- If the samples were taken from optimal points of the reservoir or taken from wrong intervals.

Because of the above points, there is a discussion if this method can be replaced by other exploration methods (such as well logging/logging while drilling LWD and downhole fluid analysis) which have recently had great developments. In brief, the validity of coring jobs can be undoubtedly trusted by overcoming its technical challenges.

2.2.2 Cost

Coring is considered a rather expensive exploration technique compared to other methods. This is a big challenge for this job. The corresponding costs are as follows:

- Cost corresponding to the rig time required to perform coring (including core-drilling, tripping, and handling);
- Rental or purchase of (sophisticated) coring tools and services;
- Rig site lab/central lab core preservation and analysis costs.

There is, however, risk for successful coring (possibility of failure in the operations, tools, etc.).

2.3   Justification

In this section, following a comparison of different exploration methods, it is investigated whether coring can be a justified exploration method.

2.3.1   Brief Comparison of Exploration Techniques

Cores are called "The Ground Truth" as they can potentially reflect the most reliable data representing the formation properties, provided that the right coring system is applied. If so, the data obtained from the core analysis can be reliably used for reservoir engineering, geological, and petrophysical studies. In order to remove the uncertainties and justify coring, a comparison between coring and other exploration methods is essentially required. The other exploration techniques are well logging, well testing, seismic survey, etc. Undoubtedly, each exploration method has some strong points and some deficiencies that should be considered. Different exploration methods contribute to complementing each other and covering their deficiencies. Thus, it should be noted that none of the formation evaluation techniques (coring, well logging, well testing, etc.) should be discarded from the outset. But rather, their possible applicability should be taken into account. The aforementioned exploration methods have been generally compared on some aspects as in Tables 2.3 and 2.4. However, some items are pointed out about their exploration value as follows:

Coring
The following points should be considered about the exploration value of coring:

- The data obtained from the rock samples (if non-damaged or with mitigated damage) can reflect the real properties of the core rock material.
- It is possible to obtain both RCAL measurements (porosity, permeability, and saturation) and SCAL measurements of the core samples.
- The data obtained from cores are normally used for calibrating other exploration methods such as well log data.
- Cores can also provide data required for rock mechanical properties. This is one of the strong points of cores, which is not obtainable by other methods.

Table 2.3 Comparison of some exploration techniques on several aspects

Aspect	Cores	Normal well logs		Well tests
		Uncalibrated	Calibrated with core data	
Porosity	✓	–	✓	–
Permeability	✓	–	–	✓
Saturation	–	–	✓	–
Relative permeability, capillary pressure, wettability	✓	–	–	–
Geomechanical properties	✓	–	✓	–
Geology (lithology, fracture studies, etc.)	✓	–	✓	–
Most reliable for exploration wells	✓	–	–	–
Order of radius of investigation	cm	m		10–100 m

Table 2.4 Approximate depth of investigation of some exploration techniques

Method	Approximate depth of invasion (m)
Coring	0.1
Well logging	0.5
DST/RFT	1–10
Well testing	50–500

- As a limiting point, the radius of investigation of cores is low. However, the obtained data can be specific to the sample.
- Core damage may occur during its drilling, retrieval, and surface handling, which requires its mitigation.

Cuttings analysis

Cuttings analysis is a popular cheap exploration method which is conducted by the rig site geologist. Some information including the lithology and sequence stratigraphy of the formations can be found by this method. However, due to the following drilling issues, its value compared to coring is questionable:

- The cuttings reach the surface in a dispersed or mixed manner. In addition, they may be affected by the drilling mud. As an example, marl might be absent in the cuttings due to low PH, but is present in the core.
- The time of sampling may not be exactly right considering the lag time for the drilling mud to transport the cuttings from the bottomhole to the surface and human error to take the samples.
- Direct geological observations of the structure, texture, mineralogical changes, faults, diagenesis, and also the estimation of rock and fluid properties (such as porosity, permeability, saturation, etc.) are not currently possible.

Well Logging

Well logging data obtained (either by wireline or logging while drilling) reflects the formation properties based on indirect measurement of geophysical properties of the waves. Well logs have a relatively larger radius of investigation than core sample and can be taken for a long reservoir interval, which is a positive point. However, some other points should be considered about this exploration method:

- Reliability of well logging data is dependent on the core data. Prior to using the log data, their calibration using actual properties obtained from core analysis data is essential; otherwise, their validity is questionable.
- Normal well logging data, in the best case, can just give an idea about the porosity and saturation values which are considered just part of data that can be obtained from Routine Core Analysis (RCAL). Thus, information about the permeability and the fluid properties are missing. To estimate permeability, costly well logging techniques such as Nuclear Magnetic Resonance (NMR) (which includes some uncertainty) are required.
- Well logging data can provide no idea of the data obtained from Special Core Analysis (SCAL), such as relative permeability, capillary pressure, wettability, etc.
- Typical well logging (via wireline or logging while drilling) by itself cannot provide reliably exact values of the rock mechanical properties (e.g., Uniaxial Compressive Strength, UCS, or Poisson's ratio). Unless, run rather costly logs such as dipole sonic (which provides P and S waves) which add up to the cost of the logging operation.

Well Testing

Well testing, pressure transient tests, particularly build-ups (after drilling a hole section) or Drill Stem Test (DST) (during or after drilling) can also be used for exploration purposes.

Using well test analysis, reliable values for the reservoir permeability, skin factor (occurred during drilling), and the reservoir thickness can be found. As the radius of investigation in well testing is considerable, the corresponding data originates from far field of the reservoir which has not undergone near wellbore mud invasion damage. However, it is a rather expensive exploration method and cannot give any idea about the most of the reservoir rock and fluid parameters.

2.3.2 Justified Coring

In order to attain a justified coring scenario, first, it is necessary to identify and consider the required reservoir data that can be found only by coring or required for calibration of measurements by the other exploration methods. In case of necessity of obtaining the core analysis data, in order to deal with the coring challenges

Table 2.5 Coring key performance indicators for a justified coring

Parameter		Definition	How
1.	Coring safety	Accounting for the risk of coring operations/tools	Highest
2.	Coring efficiency	Length of the core sample cut downhole over the length of core barrel/length of core planned	Highest
3.	Core recovery	Length of the sample recovered at the surface over the length of the sample inside the inner barrel/tube	Highest
4.	Core quality	Intactness of the core sample (no or very little damage to the core) recovered at surface. This shows itself in the accuracy of data obtained	Highest
5.	Coring cost	Rig time (the time required for cutting the sample, the tripping time, and the surface time for preparation of tools) and the cost of tools and services	Lowest Possible
6.	Coring Reliability	Functionality and the success rate in the coring jobs done by the tools	Highest

(i.e., validity and cost), it is recommended to consider the following items for a justified coring:

– Coring Key Performance Indicators (KPIs) should be considered while coring (Table 2.5). These consist of safety, coring efficiency, core recovery and quality, cost, and reliability (inferred from Briner et al. 2010; Guarisco et al. 2011; McPhee et al. 2015; Keith et al. 2016, etc.). Evaluating the KPIs helps us to evaluate the success of a coring job. Based on this, for a justified coring, it is first important that safety is never sacrificed. Second, it is significant that the maximum length of core sample is cut as planned, which is termed as maximum coring efficiency. Next, maximum coring efficiency is taken, representing maximum core column volume usable for analysis. Next, the maximum possible length of the sample should be extracted and recovered from the core barrel at the surface, which is termed as maximum core recovery. Next, highest quality core samples should be obtained. The cost of the operations should enable economic operations while using reliable tools. For specified core properties, the right selection of the coring practices and tools ensures the highest KPIs to be obtained.

– Core point/interval identification should be performed by real-time controlling of the coring operations using Logging-While-Coring (LWC) method to be placed just at the core bit. Coring automation contributes to controlling real-time core-drilling measurements and operational parameters in order to optimize the operations in a real-time manner. If the core point is missed, coring is almost unjustified.

– Consider utilizing invasion–mitigation coring techniques and systems (Chap. 7), and full-closure systems (Chap. 9) in order to, respectively, protect the core from invasion and enhance core consolidation and quality.

- Consider applying wireline continuous coring particularly in exploration wells in order to obtain higher quantity of cores so that the core analysis data could be extrapolated to the reservoir in field scale with less uncertainty. Wireline continuous coring makes numerous coring possible from multiple intervals with mitigated cost.
- Consider utilizing pressure/in situ coring systems for coring in tight rocks in order to prevent hydrocarbons particularly gas escaping from the sample while tripping and retrieval to the surface. Otherwise, tripping of tight cores can potentially cause loss of pore fluids and also mechanical damages the core.
- Apply standard handling procedures to prevent mechanical and chemical core damage at the surface.
- The last but not the least, provide sufficient efficient training for the coring personnel (Lee et al. 2013).

2.3.3 Example

In order to simply show the accuracy of the core data from a justified reliable coring operation, a typical example is presented in this section. Consider a reservoir with the planar area of 10 km^2 (2471 Acre), the thickness (h) of 100 m (328 ft), average porosity of 10%, water saturation of 10%, and oil formation volume factor (B_o) of 1.15. Using this, the Original-Oil-In-Place (OOIP) volume is computed equal to 492.08 million barrels. During the exploration phase, if the errors in the measurements of the simple parameters occur due to using uncalibrated/wrongly calibrated well log data, the errors listed in Table 2.6 would occur in the evaluation of OOIP. This signifies the importance of using reliable core data originating from a justified coring operation, to calibrate the well log data.

$$OOIP = \frac{7758A \times h \times \emptyset \times (1 - S_w)}{B_o}$$

$$OOIP = \frac{7758 \times 2471 \times 328 \times 0.1 \times (1 - 0.1)}{1.15} = 492.08 \text{ million bbl}$$

Table 2.6 Effect of errors on measurement values of Original-Oil-In-Place (OOIP)

Property	Error in measurement	OOIP (Million bbl)	OOIP Difference (Million bbl)	Typical oil price (USD/bbl)	Asset difference (Million USD)
$\emptyset$	1%	442.87	49.2	50	2460
S_w	1%	486.6 1	5.5	50	275
h	1 ft	490.58	1.5	50	75

References

Briner, A.P., A.H. Azzouni, R. Chitnis, and V. Vyas. 2010. *Sweet Success in Sour Coring*, SPE 128007 MS. Presented at the North Africa Technical Conference and Exhibition, Cairo, Egypt, Feb 14–17.

Guarisco, P., J. Meyer, R. Mathur, I. Thomson, J. Robichaux, C. Young and E. Luna. 2011. *Maximizing Core Recovery in Lower Tertiary Through Drilling Optimization Service and Intelligent Core Bit Design*, SPE/IADC 140070. Presented at the SPE/IADC Drilling Conference, Amsterdam, the Netherlands, March 1–3.

Keith, C.I., A. Safari, K.L. Aik, M. Thanasekaran and M. Farouk. 2016. *Coring Parameter optimization-The Secret to Long Cores*. Presented at the OTC Conference, Kuala Lumpur, Malaysia, March 22–25.

Lee, R.K., P.M. Strike, C.D. Rengel, and D.B. Sutto. 2013. *Harnessing Multiple Learning Styles for Training Diverse Field Personnel in Conventional Coring Operations*, IPTC 16654. Presented at International Petroleum Technology Conference, Beijing, China, 26–28 March.

McPhee, C., J. Reed, and I. Zubizarreta. 2015. *Core Analysis: A Best Practice Guide*, 52–57. Elsevier. ISBN-13: 978-0444635334.

Chapter 3
Fundamental Coring Methods

3.1 Introduction

Fundamental coring methods fall into three main categories consisting of conventional coring, wireline continuous coring, and sidewall coring. Conventional and wireline continuous methods are applied at the time of drilling and are thus named *bottomhole coring*. Sidewall coring is used after drilling. These methods are introduced in this chapter (using Anderson 1975; EXLOG/Whittaker 1985; Harrigan and Cole 2011, etc.).

3.2 Conventional Coring

Conventional coring is a method of rotary coring by which via a conventional trip, the inner tube/barrel containing the core sample is retrieved along with the outer tube assembly to the surface. As the core bit is connected to the outer tube assembly and drill pipes, conventional core-drilling is accomplished using a core bit/head with essentially the same principle as a drilling bit. However, the core bit cuts a hollow cylindrical rock and thus possesses smaller bearings and cutters than a drilling bit. The core barrel consists of an inner tube, wherein the sample enters and the outer tube (together with the overlying drill collar and jar) performs as the Bottomhole Assembly (BHA) during coring.

At the coring point, the drilling string is pulled out of the hole. Instead, the coring assembly including the core bit is lowered into the hole. After enough rock sample was cut by the bit (i.e., at the end of core-drilling), the string is pulled up suddenly (or an overpull is applied) so that the sample is broken from its bottom. A general schematic of conventional coring assembly has been shown in Fig. 3.1. The components and the method of conventional coring have been described in detail in Chaps. 4 and 5. A schematic of a core bit has been also shown in Fig. 3.2.

© Springer International Publishing AG, part of Springer Nature 2018
R. Ashena and G. Thonhauser, *Coring Methods and Systems*,
https://doi.org/10.1007/978-3-319-77733-7_3

Fig. 3.1 Schematic of
conventional coring assembly
(http://www.drillingdoc.com/
coring-technology/, accessed
on January 1, 2016)

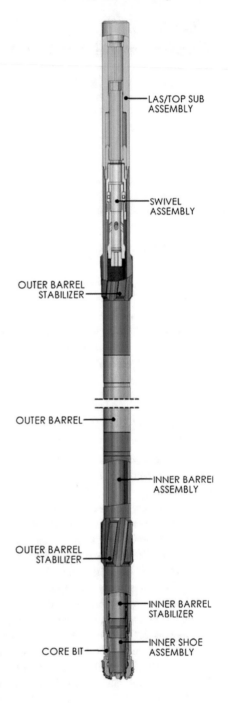

Fig. 3.2 A core bit example (http://image.china-ogpe.com/pimage/2039/image/PDC_diamond_core_bit_Business2039.jpg, accessed on September 26, 2015)

It should be noted that the most prominent advantage of conventional coring method over other methods is the possibility to cut a large size rock sample. This can be up to even greater than 5-in. diameter. Using conventional coring, the length of the core that can be cut in one run is significantly larger than the wireline method. The main disadvantage of conventional coring is that the inner tube and the sample inside can be only retrieved through a conventional drill string trip to the surface. This makes conventional coring time-consuming and thus costly.

3.3 Wireline Continuous Coring

Wireline continuous coring is another fundamental coring method that uses wireline for tripping. Thus, it does not have the disadvantage of requiring a conventional trip for each retrieval of the rock sample or pipes to the surface. This coring method is composed of two modes of drilling and coring which can be easily switched to each other using slick line/wireline. In coring mode, the inner tube assembly is lowered into the outer tube assembly by slick line. Then, mud circulation is started in order to hydraulically latch the inner tube assembly to the outer tube assembly so that the cutting of the core can commence. After core-drilling was accomplished or in case coring was terminated, the rock sample is brought up to the surface through the outer tube assembly via slick line. Then, (in the drilling mode) the inner drilling assembly including *drill bit insert/plug* at its bottom end is run in the hole again via slick line in order to proceed with drilling.

Fig. 3.3 Schematic of
wireline continuous coring
assembly

It is noted that the drilling assembly (i.e., the core bit, outer tube assembly and drill pipes) is kept in hole for both coring and drilling modes, and just the inner assemblies are intermittently changed to switch from the one mode to the other. Except for this main difference, wireline coring resembles conventional coring in other terms such as core-drilling, mud circulation path, core catching/retaining by overpull, etc. A general schematic of wireline continuous coring assembly has been shown in Fig. 3.3. Detailed explanations of the method and the required components have been explained in detail in Chap. 4 and Sect. 5.5.

Wireline continuous coring has basically mitigated the time and cost of the operation as the drilling and coring assemblies are not removed via conventional tripping. This is more substantial particularly in deephole coring and multiple zone coring when numerous cores are required. Wireline coring is highly recommended for exploration drilling when the coring points/intervals are not known in advance. Therefore, it is viable to switch from drilling to coring to obtain the rock samples at any time it is decided for.

The main disadvantage of wireline continuous coring is the rather small size of core retrieved (maximum 3 or 3½ in. in best case). This is because the cores have to be tripped to the surface through the drill pipe. Using the wireline, for each run, shorter length of core samples (e.g., 10–30 ft) can be retrieved. Therefore, the wireline method is not optimal for long coring from a single formation, but rather the conventional method is preferred.

3.4 Sidewall Coring

Particularly in exploratory wells, the formation intervals wherein obtaining cores are desired may be unidentified and be already penetrated by the drilling bit. This is termed as *missed coring points/interval*. In this case, following drilling, wireline logging is first conducted. Using the interpretations of the well logs, it is possible to approximately identify and indicate the formation interval(s) with economical or production potentials. Next, in order to verify this identification, sidewall coring method is used as the only compensating alternative to core the interval on the sides of the wellbore. Otherwise, some potentially productive hydrocarbon-bearing formations may be bypassed. Almost any time after drilling the formation, it is possible to obtain sidewall cores by dispatching sidewall coring tools downhole via wireline. The electrical wireline logging company can run the sidewall core too. As sidewall coring is less costly than conventional coring and capable of coring multiple zones, it provides a cost-effective coring method just for rough evaluations. Normally, a number of sidewall cores are taken in one run of the tool, e.g., 50 core number capacity per run.

The main disadvantages of sidewall coring are that the size of the recovered cores is small (normally maximum 1½ in. diameter and 3½ in. length); they have undergone considerable mud invasion and formation damage during coring and retrieval; and they typically lack some features interesting from reservoir

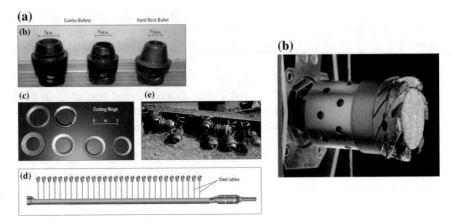

Fig. 3.4 **a** Percussion sidewall coring, and **b** rotary sidewall coring with the circular diamond core bit (Schlumberger Coring Services (2013) *Rotary Sidewall Coring—Size Matters*, Winter 2013/2014)

engineering and geological points of view particularly in fractured or heterogeneous reservoirs.

Sidewall coring falls into two main categories of percussion and rotary methods (Fig. 3.4). Percussion sidewall coring resembles perforation as explosive charges are used with just this difference that short metal tubes attached to wireline are pushed into the formation to take and hold the samples due to explosion instead of a bullet. In rotary type of sidewall coring, a rotary small bit is used to cut the plug from side of the wellbore. Rotary sidewall coring is more in practice than percussion type as it has removed some of percussion coring shortcomings, e.g., mechanical distortion of core. When the rotary bit reaches its maximum cutting depth, it is lifted upward by wireline to break the core. Then, the core is pulled back into the tool. As normally a number of cores are cut, the tool is repositioned to cut the next core.

It is noted that among the three main coring methods already discussed in this chapter, the bottom coring methods, i.e., the conventional and wireline continuous, are the main exploration techniques with significant exploratory benefits. Therefore, further discussion on sidewall coring will be out of the scope of this handbook.

References

Anderson, G. 1975. *Coring and Core Analysis Handbook*. Pennwell Corp.

EXLOG/Whittaker. 1985. *Coring Operations, Procedures for Sampling and Analysis of Bottom-Hole and Sidwall Cores*. Berlin: Springer.

Harrigan, J.J., and F.W. Cole. 2011. *Advances in Engineering and Technology of Oil Well Drilling*. USA: Koros Press Limited.

Chapter 4
Types and Components of Core Barrel Assemblies

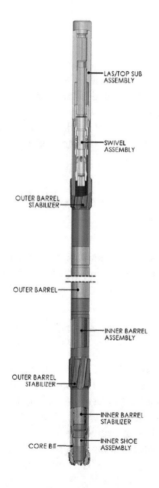

© Springer International Publishing AG, part of Springer Nature 2018
R. Ashena and G. Thonhauser, *Coring Methods and Systems*,
https://doi.org/10.1007/978-3-319-77733-7_4

4.1 Introduction

The core barrel is an essential part of each coring practice as it collects the core sample in its inner tube and simultaneously its outer tube acts as the BHA. Core barrels are not all the same, but rather differ and have different features and components depending on the coring method or system. This in turn affects their design and selection criteria.

Therefore, in this chapter, first the core barrel types and features will be discussed considering the two main bottom coring methods. Finally, the components of the core barrels and their functions are explained. This will be required for following the rest of the handbook as the prerequisite. As the core barrels used both in the petroleum and mining industries are explained, a view on the similarities and the differences of the barrels in the two industries is provided.

4.2 Core Barrel Types

Depending on the coring methods and systems to be applied, different core barrels types are used with different features and components. These can be generally categorized into *conventional, wireline continuous, slim hole, high torque, jam indicating, Jam mitigation (i.e., antijamming* and *full-closure), invasion–mitigation, oriented, pressure/in situ, Logging-While-Coring,* and *motor coring* (Table 4.1).

4.3 Components of Core Barrel Assembly (Petroleum Industry)

The components of the core barrel assemblies fall into two main categories of conventional and wireline continuous coring which are described as follows:

4.3.1 Conventional Core Barrel Assembly

The conventional core barrel assembly is typically composed of the *outer tube/ barrel (assembly)* and the *inner tube/barrel (assembly)*. The top of the outer tube is connected to the drill collars and jar (for vertical wells), or the Heavy Weight Drill Pipes (HWDPs) and jar (for directional wells). The drill collars or the HWDPs along with the outer tube assembly constitute the BHA, which provides the required weight on the bit. The bottom of the outer tube assembly is connected to the core bit. The inner tube assembly is, by itself, composed of the *upper inner tube*

Table 4.1 Main core barrel types, functions, and features

Types and systems	Function and application	Feature	Schematic
1. Conventional	For normal coring operations where the inner tube is run by a conventional string trip inside the outer tube assembly. After coring, it is retrieved again via a conventional string trip (for more information, refer to Chap. 5)	Comprises the conventional core barrel components	
2. Wireline continuous	– It is a feasible method for mitigating the rig time due to tripping (such as for coring in multiple zones, deep zones, scenarios with unclear core points, etc). It has two modes: – (Drilling mode): The wireline or slick line is used to run the inner drilling assembly (i.e., drill rods plus drill plug/ insert or pilot bit) into the hole through the drill pipes and the outer tubes. Then, it is latched to the core bit – (Coring mode): The inner tube assembly (i.e., swivel assembly, inner tubes, inner tube shoes) is run in the hole to be seated and attached to the outer tube by pumping mud or slick line/wireline (for more information, refer to Chap. 6)	– Requires wireline or slick line and overshot – Handling requires more experience and rig personnel skill	
3. Slim hole	– So far, applicable to hole diameters ranging from 4 1/8 to 4 3/4-in. – The method for slim hole coring can be conventional, wireline continuous (more	– With the same conventional core barrel components or more, but with smaller size – No safety joint is needed, but may use	

(continued)

Table 4.1 (continued)

Types and systems	Function and application	Feature	Schematic
	popular) or even coiled tubing – It is also applicable to the motor coring systems	outer tube stabilizers (if necessary)	
4. High torque (HT)	– Endurable against twist-off while make-up, drilling, or breakout It is particularly used for the following: – Long coring/core barrel runs (to 360 ft or 110 m) – Motor coring and coring in holes with high inclination angles – Coring with high Rate of Penetration (ROP) and Weight-On-Bit (WOB) – Unconventional coring (shale gas, etc.) – Suitable for *full-closure* system	– High-strength steel – Double-shouldered connections and optimized thread designs – High torsional strength	
5. Jam indicating	– It is a rod between top of the inner tubes and pressure relief plug which can be lifted in case of jam – While coring with conventional barrel, the inner tube is held in place by the mud hydraulic force. But, when jamming occurs, the friction force causes the inner tube and the jam indicator rod to be lifted, causing restriction in mudflow from ports of the inner tube plug		

(continued)

Table 4.1 (continued)

Types and systems	Function and application	Feature	Schematic
	This increases the standpipe pressure with the jam indicator. Other jam indicators are the ROP and torque decrease – In wireline continuous method, standpipe pressure instantly decreases because of the hydraulic metal-to-metal system (for more information, refer to Sect. 9.2)		
6. Antijamming[e]	– Prevents jamming and termination of coring operations in faulted or fractured carbonates, shales, unconsolidated formations, and heterogeneous sections – Increases coring efficiency and core recovery (for more information, refer to Chap. 9, Sect. 9.3)	– A high-torque core barrel is fitted with one aluminum inner tube plus two or three telescoping aluminum inner barrel sleeves – When core jamming occurs in an aluminum sleeve/tube, the shear pin connecting one sleeve to the next is sheared and enables coring and core entry to the next inner tube – May be combined with invasion–mitigation systems	
7. Full-closure	– Useful for recovering cores particularly from fractured, soft, or unconsolidated formations because (1) it enables smooth entry of core column inside the inner tube (less possibility of jamming, less core damage, and higher quality) and (2) also, it prevents loss of the sample from the bottom during Pull-Out-Of-Hole (POOH) as the full-closure catcher closes at the bottom of	The additional components required are as follows: – Two drop balls and seats are required – A slick inner tube is used for less-friction entry of the core – A dual spring-type hard-faced core catcher is used to seal the bottom of the core sample for POOH	

(continued)

Table 4.1 (continued)

Types and systems	Function and application	Feature	Schematic
	the inner tube (while tripping) – The coring mechanism resembles conventional coring with a difference: At the end of coring, when core barrel is full or if core jam occurs, the second ball is dropped to activate the *full-closure system* (for more information, refer to Chap. 9, Sect. 9.4)		
8. Invasion–mitigation	– For mitigating the extent of the mud invasion to the core sample – It consists of low-invasion, gel and sponge coring systems[d] – Useful for protecting the samples taken from high-permeable formations, and also stabilization of unconsolidated formations, etc. (for more information, refer to Chap. 7)	**Note**: Low-invasion system corresponds to the core bit and the mud properties – (Low-invasion): (1) Extended lower shoe which has the minimum diameter difference from the bit diameter, (2) using noninvasive mud – (Gel coring): High viscous non-water-soluble, noninvasive propylene glycol – (Sponge coring): using a sponge layer inside the inner sleeve to absorb oil or other fluids expelled out of the sample	
9. Oriented	– It is recommended for structurally complex reservoirs (fractured, with variable structure, etc.) – To measure hole direction (azimuth), inclination (or deviation) angle, formation dip angle, formation strike	– It is required to install *scribe knives/blades* (around a ring in lower shoe), non-magnetic drill collar (NMDC)[c], and magnetic survey equipment (e.g., gyro) – Survey tool measures the magnetic and gravity data. It is installed on the	

(continued)

Table 4.1 (continued)

Types and systems	Function and application	Feature	Schematic
	angle, fracture azimuth (or direction), azimuth (or direction) of stresses[a], Formation anisotropy, directional permeability[b] (direction of permeability), direction of fluid migration, and direction of formation deposition (for more information, refer to Chap. 10)	bearing above the inner tube	
10. Pressure/in situ	– To seal the core fluids from the downhole and at the surface accurately measure the pore fluid volumes, saturations, and estimating the initial gas/oil volume in place – For more detailed geomechanical analysis of the core rock recovered at the surface – For detailed composition analysis of the reservoir fluids (gas, oil, or water) (for more information, refer to Chap. 11)	– The core is sealed after being drilled downhole and the expanding pore fluids can rise to enter a storage canister(s) above the core in the barrel	

(continued)

Table 4.1 (continued)

Types and systems	Function and application	Feature	Schematic
11. Logging-While-Coring (LWC)	To measure the resistivity, gamma ray (GR) reading, and the dynamic drilling data (like WOB, TOB, annular temperature, pressure, and acceleration), and any other measurements (for more information, refer to Chap. 12)		
12. Motor coring	– For enhanced trajectory control of the directional and horizontal coring which requires less WOB to be applied – Proper for coring hard formations which can be fractured (for more information, refer to Chap. 13)		

[a]Sometimes, it is also called *stress orientation*
[b]In formations comprising fluvial deposits
[c]It is where the survey tool is placed
[d]Sponge Coring[TM] (Halliburton Trade Mark) and SOr[TM] (Baker Hughes GE's Trade Mark)
[e]https://vimeo.com/102673607. Accessed on September 5, 2015

assembly (or the head assembly) and the *lower inner tube assembly* (as listed and described in Table 4.2).

As described in Table 4.2, the outer tube assembly is composed of (from top to bottom) *top sub, safety joint, downhole motor* (in case of motor coring), *outer tube stabilizer, Long Distance Adjustment, LDA* (also called *Longitudinal Spacing Adjustment, LAS*), *outer tube sub, outer tubes,* and *thread protector.*

The upper inner tube assembly is composed of (from top to bottom) *LDA, swivel assembly or bearing assembly, inner tube plug, drop-ball, and seat or DAFD/drop-ball sub (in case of motor coring or wireline continuous coring).*

The lower inner tube assembly is composed of (from the top) conventional *Logging-While-Drilling sub (LWD) and its battery pack (optional), core jam indicator, inner tubes, inner tube check valve, inner tube liners, Nonrotating Inner Tube Stabilizers (NRITS), the upper shoe, the lower shoe* which includes *the core catcher,* and *the lower or bit-end bearing.*

Figures 4.1, 4.2, and 4.3 illustrate detailed schematics of some of the core barrel components including the LDA, drop-ball sub, and NRITS. Then, a detailed schematic of conventional coring has been shown in Fig. 4.4.

4.3.2   *Wireline Continuous Core Barrel Assembly*

Similar to the conventional core barrel assembly, continuous wireline core barrel assembly is typically composed of the *outer tube/barrel (assembly)* and the *inner tube/barrel (assembly).* Similar to the conventional coring method, the top of the outer tube assembly is connected to the drill collars and jar (for vertical wells), or the Heavy Weight Drill Pipes (HWDPs) and jar (for directional wells). The drill collars or the HWDPs along with the outer tube assembly constitute the BHA, which provides the required weight on the bit. The bottom of the outer tubes is connected to the core bit. However, the *continuous* term of this method indicates that there is a drilling mode in addition to the coring mode. In the drilling mode, the inner tube assembly is replaced by the *inner drilling assembly.*

Table 4.3 shows the components of the outer tube assembly (from top to bottom) which are *top sub, safety joint, locking seat and seat,* top stabilizer, outer tubes, middle stabilizer, outer tubes, and bottom stabilizer. On the bottom of the outer tube assembly (acts as part of the BHA), the core bit is connected.

For the drilling mode, a common design by some wireline coring tools providers makes the inner drilling assembly latch to the outer assembly from the top and the bottom. In accordance with this design, Table 4.3 shows the inner drilling assembly which is composed of (from top to bottom): *latch (R-mandrel, Tri-latch,* or *rope socket), flow cap, squeeze nozzle/nozzle-half, mandrel, locking grapple, float* and *float plug, drill rods, lobe sub, LWD tool & battery pack,* and *drill insert/plug* (Table 4.3). The latching is achieved on the top via the *locking grapple* and at the bottom via the *lobe sub.* The schematics of the setups for the drilling and coring modes have been illustrated, respectively, in Figs. 4.5 and 4.6. However, some

Table 4.2 Components of conventional core barrel assembly with different techniques (from top)

Main part	Components	Function	Additional info.	Schematic
1. Outer tube assembly & bit	1.1 *Top sub*	– It connects the outer tube to the drill pipes for high-torque (HT) outer tubes	– It is used for HT core barrels (for which the use of the safety joint is not necessary)	
	1.2 *Safety joint*	– For older core barrels, it was used in lieu of the top sub to connect the outer tube to the drill pipes. However, it may be used below the top sub to protect the top sub thread	– In case of the outer tube stuck, when the top sub is not used, it is designed for backing-out the inner tube assembly	
	1.3 *Downhole motor (optional)*	It is used for deviating in horizontal or directional wellbores		
	1.4 *(Outer tube) Stabilizer (s)*	– It/they keep(s) the outer tube stable – Also, it reduces the outer tube contact with the wellbore and prevents its wear and stuck and increases its outer tube(s) durability. Usually, stabilizers are ribbed (spiral) for their less contact area with the wellbore	– The number of stabilizers depends on the BHA design. For long core barrels (greater than three joints), maximum three stabilizers are used (one near-bit stabilizer, one in the middle, and the third on the top of outer tubes) and for each joint without a stabilizer, a short sub (∼ 1 m) is replaced – $L_{Stab.}$ = 3-ft – $OD_{Stab.}$ should not be less than 4 mm from D_{Hole}	
	1.5 *LDA outer tube sub*	It is ∼ 4-ft sub as an outer tube which surrounds the LDA screws	It is located below the top sub and above the top stabilizer	
	1.6 *Monel collar (just for oriented coring)*	It is a non-magnetic collar which is placed above the inner tube	The survey tool (e.g., gyro) is placed within it	

(continued)

Table 4.2 (continued)

Main part	Components	Function	Additional info.	Schematic
	1.7 *Outer tubes*	– They connect the core bit to the top of the outer tube/barrel & the drill pipes. They act as part of the BHA to transmit the rotation and the axial load/weight to the bit – High-Torque (HT) heavy-duty outer tube joints (with double-shouldered threads) are used to stand the torque (especially in highly-deviated wellbores)	– Thick-walled (0.5–1.1¼-in. thickness) stiff pipe – Consists of high tensile strength, high thickness steel outer tube joints – Tube Joint length: 30 ft (~9 m) – $OD_{O.T.} < D_{Hole}$ (0.4–1.4-in. spacing between outer tube and the hole size)[a] – Use *monel* collars (Nickel-Copper alloy, non-magnetic) for oriented coring or LWC	
	1.8 *Thread protector*	– It protects the outer tube threads from wear, etc. – It connects the *outer tubes* to the core bit		
2. Core bit/head		– It is a bit with a central hole in its center for cutting around the core column and letting it be inserted in the inner tube	– The *face-discharge* feature is used to deviate the exiting mud away from the core bit throat – Most core bits used are Poly-Crystalline-Diamond (PDC)	

(continued)

Table 4.2 (continued)

Main part	Components	Function	Additional info.	Schematic
3a. Upper inner tube assembly (i.e., head assembly)	3.1 *Long Distance Adjustment (LDA)*	– It is a locking screw (to be screwed in the locking nut) located between the inside of the top sub and the inner tube, to compensate for the thermal expansion between the aluminium inner tube and the steel outer tube Therefore, LDA helps the inner tube not to bottom-out on the core bit – Calculate[b] the distance for LDA screw accurately so that It can position the inner tubes accurately above the bit. As a result, the inner tube or liners would not expand toward the bit to be bent or broken or the core is stuck. This prevents rig time loss, especially for long barrels (compared to the *shim and sub[c]* method)	– The screw is 22-in. long (with the length 22-in./~54 cm) which allows max 11-in./~28 cm vertical movement, which has enhanced performance than the traditional shim & sub method – It is installed between the inside of the top sub and the inner tube – It is particularly required for long coring runs >90 ft and HPHT operations, or when the inner tube joints or even liners are made of aluminium	

(continued)

Table 4.2 (continued)

Main part	Components	Function	Additional info.	Schematic
	3.2 *Swivel/ bearing assembly*	– It connects the inner tube to the top sub or safety joint – It is equipped with two types of bearings (double thrust and ball) to allow the inner tubes to be stationary while the outer tube can freely rotate	– They are equipped with lubrication mechanisms to eliminate the need for surface lubricating devices (especially in HPHT) – Proper swivel guarantees great enough ROP, high core quality and recovery, and even proper mud hydraulics – The swivel assembly together with the LDA is called *Radial Longitudinal Compensation (RLC)* System	
	3.3 *Inner tube plug/pressure relief plug*	It includes *diversion ports* where the mud can exit from inside the upper inner tube assembly to the annulus between the inner and outer tubes	It is located below the swivel assembly and above the ball and seat	

(continued)

Table 4.2 (continued)

Main part	Components	Function	Additional info.	Schematic
	3.4 *Drop-ball & seat or DAFD/drop-ball sub*[d]	– Prior to the coring start, mud circulation is through the inner tube. This provides good inner tube cleaning which contributes to less subsequent core jamming or contamination. In order to begin to core, the mud circulation path should be changed to the annulus – Just prior to beginning to core, the *drop-ball* is dropped to reach and be seated in the seat above the inner tubes. This is to divert the mud circulation path from through the inner tube to the annulus between the inner and the outer tubes	– The time required for the drop-ball at the surface to reach and drop into the seat can take ½ to 1 h (depending on the mud parameters, the hole depth, and the inclination angle) – In motor coring or highly angled holes, (instead of dropping the ball from the surface), DAFD or the *drop-ball sub* is installed a few inches above the seat, as part of the inner tube assembly. It is hydraulically-activated downhole (by increasing the mud flow rate) to drop the ball into the seat	

(continued)

Table 4.2 (continued)

Main part	Components	Function	Additional info.	Schematic
3b. Lower inner tube assembly	3.5 LWD and battery (optional)	It enables real-time data (directional inclination angle, azimuth, vibration) measurement and probably transmission to the surface	Transmission of data to the surface is possible using, e.g., mud pulses	

(continued)

Table 4.2 (continued)

Main part	Components	Function	Additional info.	Schematic
	3.6 *Core jam-indicator*	– It is an inner rod and seat, located between the upper inner tube and the inner tubes, which is used to make an increase in the standpipe pressure to alert the driller in case core jamming occurs – After core jamming detection, we can pull out of the hole and thus we can prevent the core from being milled due to jamming	– Usually used in jam-prone formations such as fractured or unconsolidated rocks – A 2 ft outer tube extension sub is required in the outer tube assembly to cover it from the outside	
	3.7 *Inner tubes*	– They provide the conduit for receiving and holding the cut core column – It is possible to cover the inner surface with sponge for absorbing the expelled oil in *sponge coring*, or to fill it with *gel* for preventing static filtration and mitigating jamming tendency (especially in fractured formations), or to use *telescoping inner tubes* for anti-jamming. For more information on gel, sponge coring, and anti-jamming respectively refer to Chaps. 7 and 9	– Three types of inner tubes are available: (1) Steel (2) Chrome: it enables smooth core recovery in hard fractured formations and thus less jamming can occur (3) Disposable tubes (aluminium for temperature >140 °C, fiberglass): contribute to less mechanical damage/erosion can occur to the core and increase the coring efficiency and core recovery. They are disposed after use – Inner tubes are thin-walled ($\sim$¼-in. thick). Each inner tube joint is, e.g., 30 ft/$\sim$9 m long	

(continued)

Table 4.2 (continued)

Main part	Components	Function	Additional info.	Schematic
	3.8 *(pressure relief/vent) check valves*	– They are ball and seat valves (installed in the inner tubes prior to running the core barrel) to allow the pressurized gas in the sample to vent out of the inner tube during tripping. It relieves the pressure from inside the inner tube to: (1) prevent gas-expansion core damage during the tripping and retrieval and the surface handling, (2) provide safety during the surface handling **Note:** For special systems such as pressure coring with *triple-tubes*, check valves are installed only in the innermost inner-tube to make a passage for the gas to migrate upward and enter the mud or an overlying canister	– They have very small ball and seat mechanism – Installed for all inner tube types such as steel, aluminium, or fiberglass – They are positioned with the spacing length of 2 ft and angular spacing of zero to 90° (e.g., 25°) **Note:** In case of using triple inner tubes, the gas vents out of the whole inner tube system from its top or through the check valves of the innermost inner-tube to enter the mud or be trapped at the top of inner tube. Thus, later, the gas can be used for volume or composition measurement, etc. The oil or water driven by the gas expansion can be trapped on the bottom	 2 ft 20°
	3.9 *Inner tube liners*	– A (non-split) liner may be used inside the inner tube (steel, aluminium, etc.) to make a triple-tube system. They are used to (1) lower the friction with the core and mitigate jamming, (2) enable the expansion of unconsolidated cores especially in deephole/high temperature wells and prevent damage, (3) or in case of using systems such as pressure or sponge coring. To provide quicker and greater recovery at the surface, it may be split/cut (which is not considered a triple-tube)	– Liners are disposable and made of aluminium, fiberglass, or PVC – For pressure coring, the check valves are installed in the innermost tube and the canister is installed above. Thus, the properties of the fluid in the canister can be measured after the retrieval – Cutting or splitting of a liner may be done by laser or plasma	 Steel Inner Barrel PVC Liner

(continued)

Table 4.2 (continued)

Main part	Components	Function	Additional info.	Schematic
	3.10 *NRITS (Non-Rotating Inner Tube Stabilizers)*	– They may be installed between the inner tubes to stabilize and centralize the inner tube and prevent its wear inside the outer tube (particularly in highly-angled wells or long core barrel runs) – In addition, when breaking-out the inner tubes, there is no need for twisting the inner tubes and thus less vibration and no twisting damage can occur, which contributes to enhanced core recovery and quality	– It is recommended for long/extended core runs >30 ft (>~9 m) and high-torque (HT) core barrels – While tripping-in, they are joined to each inner tube joint – At the surface, the inner tube joints will be easily broken by hydraulically-driven guillotine blade into a window in NRITS	
	3.11 *Upper shoe*	An extension which connects the lower shoe to the inner tube (to prevent the inner tube thread damage)		
	3.12 *Lower shoe*	– It is the bottom end sub of the inner tube assembly, which holds the core catcher and seats into the bit throat – It acts as a guide for the core to enter and to stabilize the lower end of the inner tube against the core bit	– It should be ~8–13 mm apart from the bit inner shoulder – Two types: (1) Conventional lower/pilot shoe, and (2) extended low-invasion shoe (which extends nearer to the bit and thus reduces the mud invasion prior to entry of the sample into the barrel, as a feature of low-invasion systems)	 Conventional/pilot (left), Extended low-invasion (right)
	3.13 *Bit end/lower bearing*	It fits into the bit shoulder as an additional measure to ensure the inner tube bottom not to rotate with the outer tube		

(continued)

Table 4.2 (continued)

Main part	Components	Function	Additional info.	Schematic
	3.14 Core catcher/ breaker/dog	– It is installed inside the lower shoe to grab and break the core at the bit throat after the overpull (using its tungsten-carbide hard-faced grits with 5° locking taper) – Full-closure catchers[e] can be used for coring in unconsolidated formations, such as fractured unconsolidated where core jamming can occur especially at the catcher	– The ID of the core catcher is surfaced with tungsten carbide – There are different types: (1) Spring (which exerts a side force like Chinese finger cuffs), or (2) Flapper (which closes under the weight of the core)	

[a]This spacing is designed to allow possible fishing of the outer tubes (refer to Chap. 15)

[b]Equal to the measured gap left in top sub (after screwing top sub in cartridge bowl in the outer tube swivel assembly (Fig. 4.1) plus the relative expansion of the aluminum or fiberglass inner tubes relative to the steel ones

[c]In the shim and sub method, traditionally, some spacing shims were used on the top sub for compensating for the thermal expansion differential. This method could only provide around 20% of thermal compensation of that with the LDA method

[d]DAFD stands for "Downhole Activated Flow Diverter". It is also called "fast ball sub" by Reservoir Group

[e]Owing to its slick entry and hydraulic overpull, it contributes to higher core quality and efficiency. More information on the full-closure system can be found in Chap. 9 (Sect. 9.3)

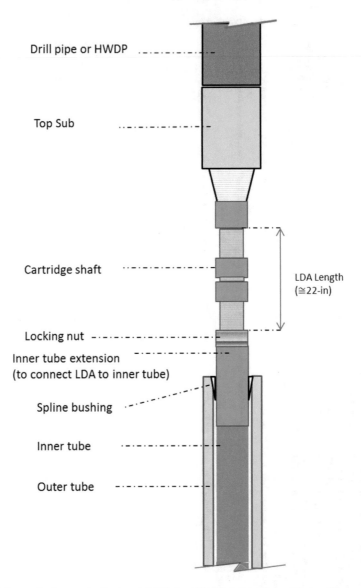

Fig. 4.1 Long displacement adjustment, LDA (modified from courtesy of Baker Hughes GE)

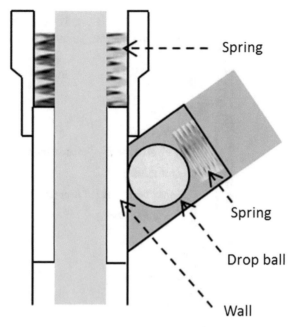

Fig. 4.2 *Drop-ball sub*, which includes a ball kept in place in front of a restored spring and behind a wall (yellow), which can be removed by a pressure surge. It is used with motor coring or highly deviated wellbore where dropping the ball from the surface is not possible

Fig. 4.3 Nonrotating inner tube stabilizer (NRITS). **a** Schematic of two halves of NRITS, **b** using *torque lock* to break out the inner tubes and core cutting by the *splitter*, and **c** two halves of NRITS at the end of breaking out (published courtesy of Baker Hughes GE)

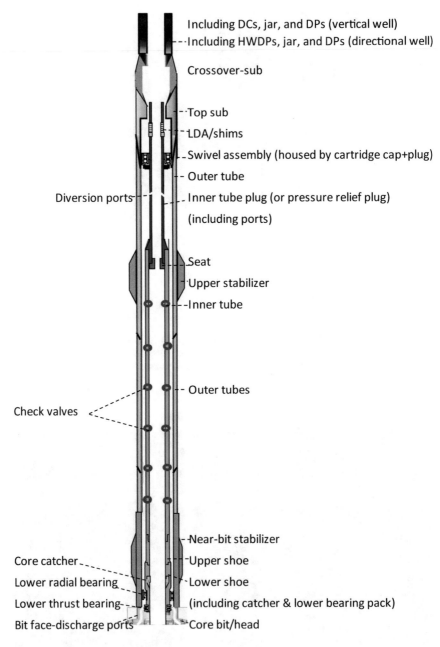

Fig. 4.4 Schematic of a typical conventional core barrel

Table 4.3 Components of wireline continuous coring and drilling assemblies (from the top)

	Components	Function	Schematic
		1. Outer tube assembly	
1.1	*Top sub*	It connects the outer tube to the drill string (for HT barrels)	
1.2	*Safety joint*	It connects the outer tube to the drill string in lieu of the top sub or to protect the top sub threads	
1.3	*Locking/landing seat*[a]	It is the place where the *grapple* locks or is seated	Locking seat / Seat
1.4	*Outer tube joints*	They compose the string or conduit and provide the weight and stability either for the coring or drilling mode	

able 4.3 (continued)

	Components	Function	Schematic
1.5	Top stabilizer	It provides the stability (above the outer tubes)	
1.6	Middle stabilizer	It provides the stability (in the middle of the outer tubes)	
1.7	Bottom stabilizer	It provides the stability (at the bottom of the outer tubes)	
		2. Core bit/head	
		It is used for cutting a cylindrical core of rock	
		3. Drilling mode (inner drilling assembly) for a common set-up	
3.1	Latch (for overshot)[b]	It is used to be attached and gripped by the overshot for Run-In-Hole (RIH) or Pull-Out-Of-Hole (POOH) of the assembly	
3.2	Flow cap with nozzles	It includes some nozzles in order to direct the mud flow (from above) into the inner drill rod. This creates a jetting force downward	
3.3	Squeeze nozzle/Nozzle-half	It provides the required downward mud jetting force and thrust on the mandrel to tighten the *locking grapple* on the *locking seat*	
3.4	Mandrel	It is used to apply thrust on the *locking grapple* to lock it in place or to unlock it prior to the wireline retrieval by removing the thrust on the mandrel	

(continued)

Table 4.3 (continued)

	Components	Function	Schematic
3.5	*Locking/landing grapple* (for latching or tightening the inner tube assembly from the top)	It is seated and locked into the *locking seat* (as a mechanism of keeping the inner drilling assembly tightly in place and withstanding upward force due to the WOB)	
3.6	*Float & float plug*	They are used for well control reasons while running the inner drilling assembly with the outer tube in the hole	
3.7	*Drill rods*	They provide WOB for the drill plug **Note:** In wireline continuous coring, the drill rods and generally the inner drilling assembly is run by the wireline or slick-line	
3.8	*Lobe sub (for also hydraulically latching from the bottom)*	– It is used as an auxiliary locking system which is hydraulic: it latches the inner drilling assembly to the outer tube hydraulically (hydraulic metal-to-metal seal) – In order to unlatch at the end of coring, it is just necessary to stop the mud circulation	
3.9	*LWD sub & battery pack* (Optional)	– It can be installed above the core barrel for data measurement and transmission to the surface – The data measured real-time data consist of: directional inclination angle, azimuth, vibration, GR, automatic inclination hold), and reservoir evaluation data (porosity, resistivity, etc.	
3.10	*Drill plug/insert*	– It composes the inner bit part (i.e., which is inserted in the core bit) to perform drilling – $D_{insert} = D_{core}$ (e.g. 3 or 3½-in.)	

(continued)

Table 4.3 (continued)

	Components	Function	Schematic
		– Depending on the design of the wireline coring tool (company design), it can have, e.g., two nozzles with the size of 10–16/32-in. for the mud to flow through or may have no nozzles and thus no mud flow through it	
		4. Coring mode (inner tube assembly)	
		4a. Upper inner tube assembly	
4.1	Latch (for overshot)	It is used to be gripped by the overshot for RIH or POOH of the assembly. The overshot is connected down the slick-line/wireline and latched to the latch	
4.2	Flow cap with nozzles	Flow cap includes some nozzles in order to direct the mud flow from above into the inner tube with jetting force	
4.3	Pressure head assembly (with squeeze nozzle)	They provide the required hydraulic mud jetting force or required thrust to keep the inner tube down seated on the seat by the hydraulic power Note: The pressure head in the coring mode has replaced the mandrel in the drilling mode	
4.4	Cartridge cap & plug	They provide the housing for the bearing assembly	

(continued)

Table 4.3 (continued)

	Components	Function	Schematic
4.5	*LDA/shims*	It is used to compensate for the thermal expansion and for the longitudinal compensation of the inner tube assembly	
4.6	*Swivel assembly/radial bearing assembly*	It prevents the rotation of the inner tube at the upper point of contact with the outer tube (radial compensation)	
4.7	*Inner tube plug/pressure relief plug*	It includes a few ports where the mud can exit from inside the upper inner tube (or the head assembly) to the annulus between the inner and the outer tubes	
4.8	*Drop-ball sub or DAFD*	– The conventional drop-ball (to be dropped from the surface) is not possible to be used in wireline continuous coring, motor coring, etc. – The ball is hydraulically activated by the mud pressure surge to be dropped from the drop-ball sub pr DAFD on the seat for diverting the mud to the annulus between the inner and outer tubes	

(continued)

Table 4.3 (continued)

	Components	Function	Schematic
4.9	LWD tool & battery pack (Optional)	It enables real-time measurement and transmission of data (directional inclination, azimuth, vibration, etc.) to the surface	
		4b. Lower inner tube assembly	
4.10	Inner tubes	They provide the conduit for holding the cut core column **Note**: In wireline continuous coring, the inner tubes may be pumped[c] in place by mud instead of using the slick-line/wireline	
4.11	NRITS (Non-Rotating Inner Tube Stabilizers)	– They are placed between the inner tubes to stabilize, centralize, and prevent the wear of the inner tubes inside outer tube (particularly in highly-angled wells or long core barrel runs)	
4.12	Upper shoe	It is an extension that connects the inner tubes to the lower shoe	

(continued)

Table 4.3 (continued)

	Components	Function	Schematic
4.13	*Lower shoe*	It is the bottom-end sub of the inner tube assembly which holds the catcher and seats into the bit throat to act as a guide for the core to enter and to stabilize the lower end of the inner tube against the core bit	Pilot (left), Extended (right)
4.14	*Bit-end/lower Bearing*	It fits into the *bit landing shoulder* as an additional measure to ensure that the inner tube does not rotate with outer tube/barrel. It also contributes to internally centralizing and stabilizing the inner tube assembly	
4.15	*Core catcher*	– It is placed inside the lower shoe to grab and break the core at the bit throat while overpull using its tungsten-carbide hard-faced grits with 5° locking taper	

[a]It is used when the latch assembly is designed to be located on the top of the outer tube assembly (for tools provided by companies like NOV)

[b]Also called "R-mandrel", "triple latch assembly", or "rope socket" (*depending on the tool or company terminology*)

[c]As used by the mining industry companies and few petroleum industry companies

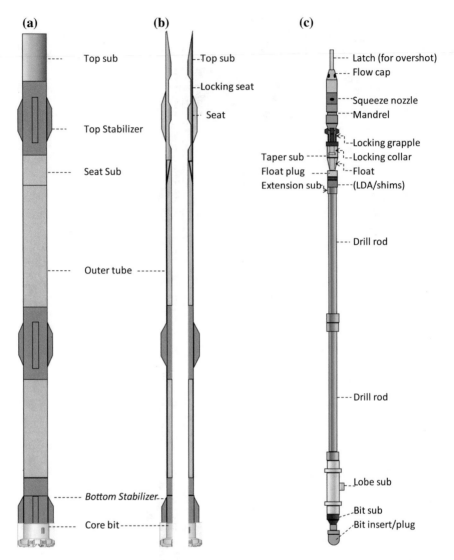

Fig. 4.5 The drilling mode assembly for a common type of wireline continuous method, **a** the overall assembly for the drilling mode, **b** the drilling BHA including the outer tube assembly, and **c** the inner drilling assembly (inferred partly from Ahmed et al. 2013 and Farese et al. 2013)

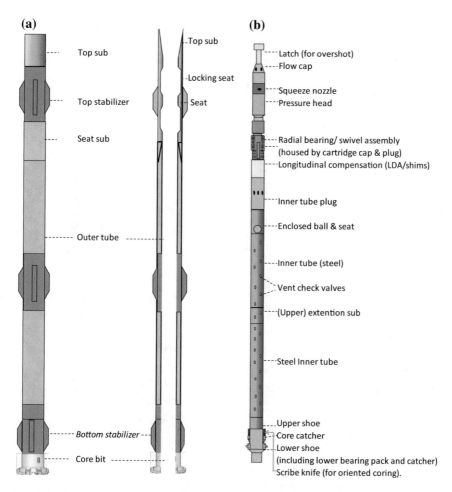

Fig. 4.6 The coring mode assembly for a common type of wireline continuous method, **a** the overall coring assembly including the outer tube assembly, and **b** the inner tube assembly (inferred partly from Ahmed et al. 2013 and Farese et al. 2013)

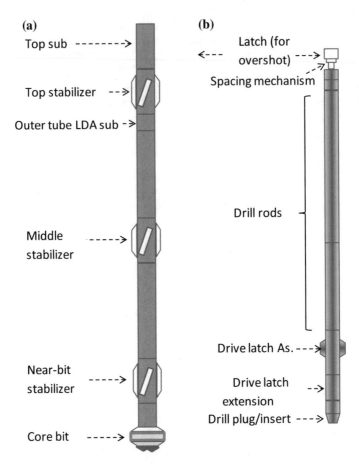

Fig. 4.7 The schematic for another type of continuous wireline method: **a** the outer tube assembly, and **b** the inner drilling assembly (drilling mode)

others' designs make latching in the drilling mode possible only from the bottom using *drive latch assembly* as shown in Fig. 4.7. There are some minor differences in the second design compared with the first one; for example, for the latching, the second design uses the *drive latch assembly* (instead of the *locking grapple* and *lobe sub*) to make the drill rods unison with the outer tube and the bit insert/plug in the bottom so that they can rotate with the same rate as the outer tube (Fig. 4.8).

For the coring mode, the inner tube assembly is, by itself, composed of the *upper inner tube assembly* (also called the *head assembly*), and the *lower inner tube assembly*. The upper inner tube assembly is composed of (from top to bottom) *R-mandrel, flow cap, pressure head assembly, cartridge cap and plug, swivel or bearing assembly, inner tube plug, and drop-ball sub* (Table 4.3). The lower inner tube assembly is composed of (from top) *LWD and battery pack (optional), inner tubes (including check valve in walls or optionally inner tube liners), NRITS (nonrotating inner tube stabilizers), upper shoe,* and the *lower shoe* which includes

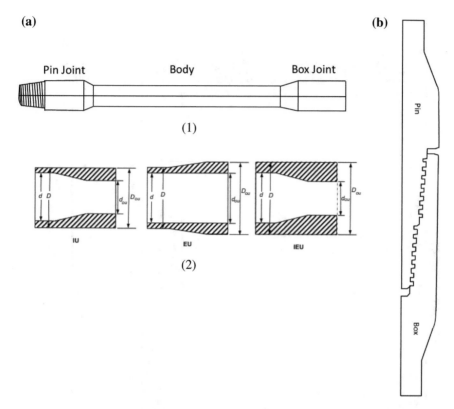

Fig. 4.8 Schematics of **a** API drill pipes: (1) components including the tool joints, (2) with internal upset, IU, external upset, EU, and internal-external upset, IEU (Mitchell and Miska 2011), and **b** half-section of typical mining drill rods (which has only minimal internal-upsets)

the *lower/bit-end bearings* and the *core catcher*. It is noted that as jamming can be instantly detected because of the hydraulic seating on the pressure head metal-to-metal seal, no *core jam indicator* is used (refer to Chap. 6, Sect. 9.2.2).

4.4 Components of Core Barrel Assembly (Mining Industry)

The core barrels used in mining industry follow the wireline continuous method and similar to the system used in the petroleum industry consist of two main parts:

1. The outer tube assembly,
2. The inner tube assembly (which consists of the head assembly and the lower inner tube).

Table 4.4 Components of the core barrel assembly used in wireline continuous coring (mining industry)

Section	Component	Function	Special features: reason	Schematic
1. Outer tube assembly	1.1 *Locking coupling*	The pin-end contracts the opened latching system to keep the inner tube assembly locked/latched in place (i.e., the inner tube in place at the bottom of the hole during coring)	It is heat treated for additional wear resistance	
	1.2 *Adaptor coupling*	It allows the latching system to open fully and locks the *landing ring* inside the box end of the outer tube		
	1.3 *Landing ring*	It is positioned inside the outer tube to ensure the landing of the inner tube assembly in the outer tube (i.e., the inner tube landing shoulder sits on landing ring with a landing indicator)	It is heat treated and has resistance against impact when the inner tube seats	
	1.4 *(Back) Reamer*	It is used for enlarging the wellbore or reaming the tight wellbore sections	Some PDC cutters are used on gauge	

(continued)

Table 4.4 (continued)

Section	Component	Function	Special features: reason	Schematic
	1.5 *Outer tubes*	– The coring BHA (in mining) constitutes the outer tube joints for transmitting the rotary speed and WOB to the bit – The length of each joint is 1.5 or 3 m and the total length can be 1.5, 3, 6, and 9 m[a] – Unlike petroleum coring, stabilizers are not needed – The pin end connects to the inner tube stabilizer	The material is from stiff steel which can withstand great loads	
	1.6 *Inner tube stabilizer*	– It keeps the inner tube stable and centralized inside the bit for receiving the core sample	The drilling mud can pass through it to make the bit lifelong	
	1.7 *Thread protector*	– It is used for protecting the threads of especially new outer tubes	It has longer outer tube threads life	
	1.8 *Reamer*	It is used for enlarging the hole or preventing the tight hole of wellbore walls	It is equipped with PDC cutters on gauge	

(continued)

Table 4.4 (continued)

Section	Component	Function	Special features: reason	Schematic
2. Upper inner tube assembly (i.e., head assembly) 2.1 Upper head assembly[b]	2.1.1 *Spearhead*	The tip of the upper inner tube assembly which is latched to the overshot for run in the hole or pull out of the hole of the inner tube assembly		
	2.1.2 *Latch retracting case*	– It covers the upper latch body – In older systems, it used to mechanically release the latches from the locking coupling		

(continued)

Table 4.4 (continued)

Section	Component	Function	Special features: reason	Schematic
	2.1.3 Latches/latch assembly	– It opens the inside adapter coupling and pushes against the face of the locking coupling to lock the inner tube inside the outer tube – It keeps the inner tube in position and prevents its movement with the core entry		
	2.1.4 Upper latch body	It accommodates the latching mechanism and makes the upper part of the drilling mud bypass		
	2.1.5 Landing shoulder	– It is a ring between the upper and lower latch bodies – It makes contact with the landing ring from the outer tube components to prevent the inner tube assembly from traveling completely through the outer tube assembly and forms a barrier which forces the drilling fluid to travel through the ports in the upper and lower latch bodies which is key to landing indication	Heat treated for greater resistance	

(continued)

Table 4.4 (continued)

Section	Component	Function	Special features: reason	Schematic
2. Upper inner tube assembly 2.2 Lower head assembly	2.2.1 *Lower latch body*	– It holds the landing shoulder in position by threading into the upper latch body – If landing indication is used, it accommodates the steel ball and bushing. It makes the lower part of the drilling mud bypass		
	2.2.2 *Spindle and lock nut*	It is the connector between the upper and the lower parts of the head assembly and provides length changes for the inner tube (i.e., adjusting the inner tube gap) depending on the formation strength	For example, in soft formations, the inner tube gap from the bit should be smaller to prevent the mud from flushing away	
	2.2.3 *Shut-off valves*	– They are the rubber disks placed on the spindle used for indication of filling of the inner tube or core jam – When the inner tube resists further core entry (due to jamming or the inner tube being filled), washers are compressed to expand and thus increase the surface mud pressure and alert the driller	It must have flexibility and expanding property due to the inner tube upward movement and resultant thrust loading	
	2.2.4 *Valve adjusting washers*	They are metal washers that work in combination with the shut-off valves for core jamming indication		
	2.2.5 *Thrust bearings*	They allow the upper head assembly to rotate freely while the lower head assembly and inner tube can stay stationary (i.e., nonrotated)		

(continued)

Table 4.4 (continued)

Section	Component	Function	Special features: reason	Schematic
	2.2.6 Spindle bushing/bearing	It connects the bearing assembly and the inner tube cap assembly		
	2.2.7 Compression spring	As the drill string is pulled back during the core retrieval process, its compression transfers enough force from the inner tube to the bit to break the core		
	2.2.8 Inner tube cap assembly	It includes the ball and seat. During run in the hole, the ball is pushed up to allow the mud pass through the inner tube. Just before coring, the ball is seated in its seat to change the mud path to the annulus between the inner and outer tubes to prevent the core flushing		
3. Lower inner tube assembly	3.1 Inner tubes	– It receives the core column – The joint length can be 1.5, 3, or 6 m **Note**: For soft formations, avoid tapping by hammers; instead, use triple inner tubes (with the innermost liner which can be opened into two halves)	The wall of the inner tubes must be smooth to prevent the sticking of the core sample to the tube walls	

(continued)

Table 4.4 (continued)

Section	Component	Function	Special features: reason	Schematic
	3.2 *Stop ring*	It is locked into a groove inside core lifter case to keep the core lifter in place	It is the first piece of the check valve system which prevents the cut core from dropping out of the inner tube during retrieval	
	3.3 *Core lifter*	It is designed to let the core sample move easily up into the inner tube but when the core is pulled back, the core lifter grips the sample which results in it being pulled down further into the taper of the core lifter case. This action forces the slot in the core lifter to close even tighter onto the sample ensuring it does not slip as it is being retrieved out of the hole	It is the second piece of the check valve system	
	3.4 *Core lifter case*	Working in combination with the compression spring, it forces core lifter downward to grip and break the core tightly	It is the second piece of the check valve system	

Modified from courtesy of Sandvik (2013)

[a]The core barrel length of 1.5 m is only used for underground and heli-portable operations. For soft fragile formations (like coal), the overall length of 3 m is in practice (with H or P size). For harder or deep formations, inner tubes or core barrels with the overall length of 6 and 9 m are used, which are called extended. 9 m is just rarely practiced. The length of the core barrel and the inner tubes depends on the mast height (space limitation) and the safety concerns

[b]This is the most complex part of the inner tube which enables latching, landing indication, and s the mud circulation to the bit

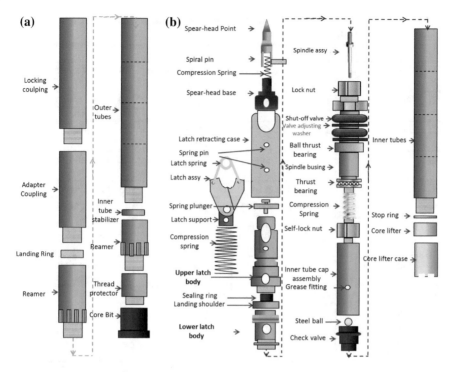

Fig. 4.9 The whole mining core barrel components. **a** The outer tube assembly, and **b** the inner tube assembly comprising the upper inner tube assembly (the upper head assembly and lower head assembly), and the lower inner tube assembly (from top to bottom)

The components of the outer and the inner tube assemblies are explained and illustrated in Table 4.4 and Fig. 4.9 (from top to bottom). The core barrels used in the mining industry have some similarities and differences from the ones in the petroleum industry. First, they have no counterparts with the petroleum conventional coring tools. But rather, they are quite similar to the wireline coring tools used in the petroleum industry (already discussed in Sect. 4.3.2) with rather different terminologies. This originates from the fact that the latter was originally developed from the first, but just has been translated and updated to meet the petroleum coring and drilling traditional practices (refer to Sect. 5.1).

The differences in the two industries' wireline coring tools are as follows: First, in petroleum practices, both coring and drilling modes are available which can alternate intermittently, whereas in mining practices, no drilling mode exists and thus no inner drilling assembly/tool is available. Second, in mining, the pipes used above the core barrel (called drill rods) through which the core samples are retrieved via wireline are different from the API drill pipes used in the petroleum industry (look at Fig. 4.8). This constitutes the main origin of the differences of the coring systems used in both industries (including the core barrel dimensions and configuration, the surface handling tools, and the obtainable core size) (Tables 4.5 and 4.6).

Table 4.5 Legend to Fig. 4.10 consisting of the core barrel and the head assembly components (courtesy of Sandvik 2013)

(a)		(b)						
1.	Head assembly	1.	Spearhead point	10.	Coiled spring pin	19.	Shut-off valve
2.	Inner tube	2.	Compression spring	11.	Upper latch body	20.	Valve adjusting washer
3.	Stop ring	3.	Detent plunger	12.	Landing shoulder	21.	Bearing
4.	Core lifter	4.	Coiled spring pin	13.	(Steel) Ball	22.	Spindle bushing
5.	Core lifter case	5.	Spearhead base	14.	Bushing	23.	Bearing
6.	Locking coupling	6.	Latch retracting case	15.	Lower latch body	24.	Compression spring
7.	Adaptor coupling	7.	Spring pin	16.	Nord-lock washer	25.	Self-locking nut
8.	Landing ring	8.	Latch spring	17.	Lock nut	26.	Inner tube cap assembly
9.	Outer tubes	9.	Latch	18.	Spindle		
10.	Inner tube stabilizer						
11.	Thread protector						

Table 4.6 Legend to Fig. 4.11. Surface-underground core barrel and the head assembly components (Courtesy of Sandvik 2013, HSU)

Core barrel assembly components (a)

1.	Head assembly	4.	Core lifter H	7.	Adaptor coupling	10.	Inner tube stabilizer
2.	Inner tube	5.	Core lifter case	8.	Landing ring	11.	Thread protector
3.	Stop ring H	6.	Locking coupling	9.	Outer tube		

HSU head assembly components (b)

1.	Latch body	11.	Spring	21.	Washer	31.	Pin
2.	Latch HSU	12.	Release tube	22.	Upper bearing housing	32.	Ball
3.	Spring pin	13.	Shaft	23.	Compressions spring	33.	Check valve body
4.	Spring pin ext.	14.	Valve body	24.	Bushing	34.	Inner tube connection
5.	O-ring	15.	Nut	25.	Ball bearing	35.	Spacer tube

(continued)

Table 4.6 (continued)

Core barrel assembly components (a)							
6.	Stop ring	16.	Shut-off valve	26.	Washer	36.	Spacer tube
7.	Landing shoulder	17.	Washer	27.	Nut-lock-king	37.	LS spacer
8.	Release valve	18.	Spacer washer	28.	Lower bearing housing	38.	Stabilizing tube
9.	Socket set screw	19.	Thrust bearing	29.	Grease fitting	39.	Spring pin
10.	Pin	20.	Ball bearing	30.	Nut	40.	O-ring

In petroleum coring and drilling, the bodies of the drill pipes are relatively small-sized compared with the hole (because they commonly have external-upset tool joints as shown in Fig. 4.8a), and thus there is a relatively large clearance between the hole and the body of the drill pipes (e.g., 44.5 mm, using 5-in. drill pipes in a 8½-in. hole). In contrast, the mining drill rods are uniformly sized (i.e., without external tool joints, but with small internally upset tool joints) and thus the clearance between the hole and the mining drill rods is minimal (2–5 mm, as shown by δ in tables of Sect. 15.2.2). In addition, the mining drill rods are significantly thinner (less thickness) than the API drill pipes. Therefore, for a specified hole size, the ID of the mining drill rods is significantly larger than that of petroleum drill pipes. This signifies the fact that samples to be retrieved via wireline through the mining drill rods do not suffer from the same core size limitation as in the petroleum coring. Next, the threads of the pipes are different. Therefore, the coring tools used in the two industries differ in size and threads. There are some discussions about accelerating an exchange of the coring tools and technologies between the two industries (particularly from mining to the petroleum industry, as discussed by Ashena et al. 2016). Particularly, in slim holes of 4 3/4-in. and smaller, using the petroleum wireline coring, we can hardly obtain core sizes large enough for proper lab analysis (with the minimum diameter size of 1.85-in.), whereas using mining style coring tools, we can obtain large enough core samples (as an example of the system, refer to Table 15.10).

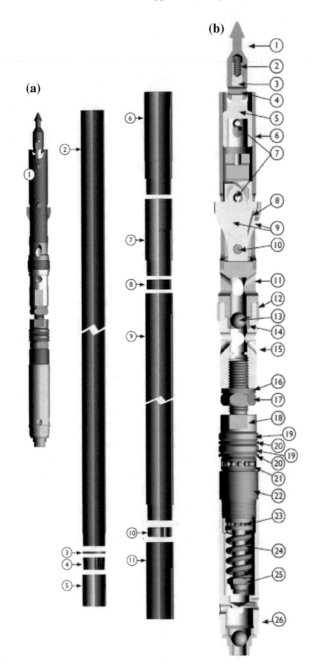

Fig. 4.10 Schematics of **a** the core barrel assembly (with the inner tube on the left, and the outer tube on the right), and **b** the head assembly (courtesy of Sandvik 2013)

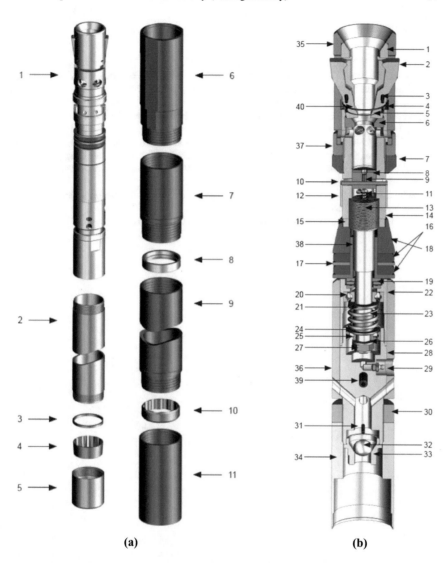

(a)   **(b)**

Fig. 4.11 The schematic of surface-underground coring components: **a** the core barrel assembly with the inner tube (left) and the outer tube (right), and **b** the head assembly. Depending on the type of the drill rods, the core barrel can be HSU, NSU, etc. [courtesy of Sandvik (2013), Sandvik HSU, a modern core barrel recently developed which is appropriate for both underground and surface coring operations (SU)]

References

Ahmed, H., T. Farese, A. Mohanna, I. Adebiyi, and O. Al-faraj. 2013. *Coring Optimization: Wireline Recovery Using Standard Drill Pipe*, SPE 166739-MS. Presented in SPE Middle East Drilling Technology Conference & Exhibition, Dubai, October 7–9.

Ashena, R., G. Thonhauser, W. Vortisch, H. Ott, A. Elmgerbi, C. Gruber, and A. Roohi. 2016. *Preserving Shale Gas Core Quality during Tripping*. Presented in the 79th EAGE Conference and Exhibition (Annual) 2017, Paris, France.

Farese, T., H. Ahmed, and A. Mohanna. 2013. *A New Standard in Wireline Coring: Recovering Larger Diameter Wireline Core Through Standard Drill Pipe and Custom Large Bore Jar*, SPE 163507. Presented at the SPE/IADC Drilling Conference and Exhibition, Amsterdam, The Netherlands, March 5–7.

Mitchell, R.F., and S. Z. Miska. 2011. *Fundamentals of Drilling Engineering*. SPE Textbook Series No. 12, ISBN 978-1-55563-207-6.

Sandvik, H. 2013. Core Drilling Handbook. '*Reaching Greater Depths*'.

Chapter 5
Conventional Coring

5.1 Introduction

As mentioned in Chap. 3, conventional coring is a traditional method of taking samples from bottom-hole formations. This method is still the most commonly practiced method in the petroleum industry whereas it is almost never used in the mining industry.

In this chapter, following a description of the conventional coring including its schematics, its step-by-step procedure is presented. Finally, the coring challenges and particularly the conventional coring challenge are discussed.

5.2 Description

Based on Sect. 3.2, the basis of conventional coring method is that the core barrel assembly (comprising the inner and outer tube assemblies) is attached to the conventional drill string/pipes. Therefore, at the coring point, first the drilling assembly must be pulled out of the hole. Next, the core barrel assembly is run in the hole via a conventional trip in order to be positioned downhole for starting coring.

Following positioning of the core barrel bottom hole and prior to coring begin, mud circulation is established through the inner tube to ensure it is clear off any debris or junk prior to the sample entry. After coring commences, the mud circulation path is diverted to the annulus between the inner and outer tubes in order not to flush and damage the cut core. This is done by virtue of a *drop-ball* which is either dropped from the surface to be seated on the *seat*, or it is already installed on

© Springer International Publishing AG, part of Springer Nature 2018
R. Ashena and G. Thonhauser, *Coring Methods and Systems*,
https://doi.org/10.1007/978-3-319-77733-7_5

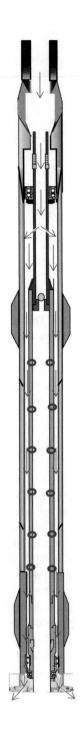

the side of the core barrel and is dropped by hydraulic activation. It is noted that this system of circulation path change is possible only when the conventional core barrel is simple or initially open.[1]

When the coring starts and the core sample is being cut, it enters the inner tube. During coring, the inner tube is kept stationary while the outer tube is rotating with the core bit.[2] After enough rock sample was cut by the core bit (i.e., the end of core-drilling), the string is suddenly pulled up or overpulled so that the sample is cut from bottom by virtue of *core catchers* or *retainers* at the bottom end of the inner tube. At the end of coring, the coring string is conventionally pulled out of the hole.

The conventional core barrel or its components (discussed in Sect. 4.3.1), is very critical for a successful coring job. This signifies its right selection. The primary parameters to be considered are (1) the hole size to be cored, and (2) the required core size (depending on the coring objectives). The parameters and specifications of some currently available conventional core barrels have been presented in Chap. 15.

The schematics of conventional coring prior to coring (i.e., before the ball is dropped) and while core-drilling (i.e., after the ball was dropped) have been, respectively, shown in Figs. 5.1 and 5.2. The red arrows represent the mud flow path.

5.3 Procedure

The procedure for conventional coring is as follows:

1. Make sure that the coring and core handling-related personnel are adequately trained.
2. Check the BHA (the outer tubes, stabilizers, drill collars, and jar) and the inner tube assembly (i.e., the inner tube, the swivel assembly, float, etc.) carefully before running in hole.
3. Pick up the first (bottom) outer tube joint (from mouse hole) using *lift bail* and *elevator* and connect it to the stabilizer, lower it into the Polycrystalline Diamond (PDC) bit, run-in-hole, make as many outer tube connections as necessary, set the slips and clamps (Fig. 5.3).

---

[1]Otherwise if the barrel is combined with gel or sponge, etc., only closed-end inner tubes can be used. In such inner tubes, the ball is already seated in the seat and thus the inner tube is closed from its top. This makes the mud circulation through the inner tube impossible. Therefore, before tripping the drill string to the core point, care must be taken to wash and condition the bottom hole.

[2]Traditionally, the conventional coring was conducted with both the outer and inner tubes rotating. This caused mechanical and vibration damage to the recovered rock sample in the inner tube. As a primary innovation in the core barrels, bearings/swivels were added to the inner tube assemblies so that the outer tube assembly is free to rotate whereas the inner tube can remain stationary.

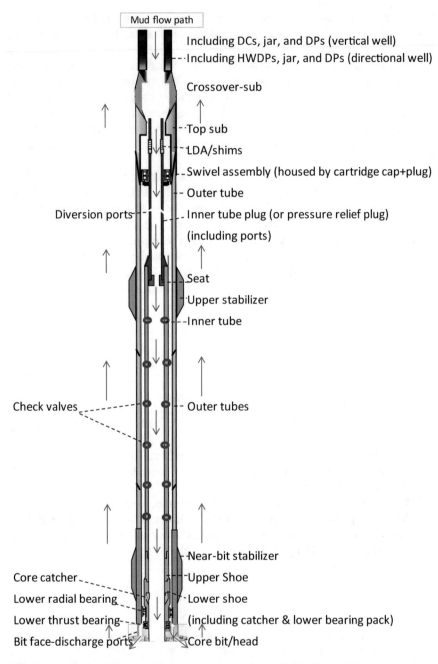

Fig. 5.1 Conventional coring prior to coring start or drop of the ball

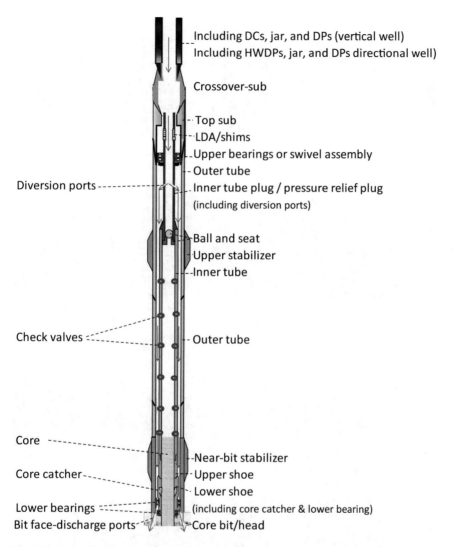

Fig. 5.2 Conventional coring after the ball was dropped (during coring), illustrating conventional core barrel and mud circulation path shown by red arrows

(a) Assembling and carrying the outer tube near the catwalk

(b) Using *tugger line* and then elevator to carry the outer tube to the floor

(c) Hold the outer tube near the rotary table, disconnect the protector bottom sub

(d) Make the core bit ready

(e) Connect the bit to the stabilizer and the outer tube

(f) Run-In-Hole (RIH)

(g) Use rig tongs to apply torque

(h) RIH the outer tube assembly

Fig. 5.3 The BHA assembly in conventional coring (using the top-drive system). The connection joints of the outer and the inner tubes have been already made on the ground for quicker operations (published courtesy of Baker Hughes GE)

(i) Set slips and clamps around the
outer tube

(j) Disconnect the top joint of the
outer tube

(k) Lifting the inner tube from the ground
through the catwalk to the rig floor

(l) The inner tube with protector is
on-the floor

(m) Loading the outer tube with the
inner tube using the tugger line

(n) Top of the inner tube

(o) Using clamps to take inner tube on the
outer tube and letting the *lift bail* to be released

(p) Replacing the *lift bail* with the *top sub*

Fig. 5.3 (continued)

(q) Releasing the clamps around the inner tube and ready for connecting *top sub* with the outer tube joint

(r) Lower Top Sub into the outer tube joint and ready for connecting them

(s) Releasing the clamps around the outer tube and slips

(t) Checking the bit nozzles and its throat and getting ready for RIH

Fig. 5.3 (continued)

4. Pick up the inner tube assembly from the ground to the rig floor using *tugger line*,[3] thread the inner tube shoe assembly (including core catcher) to the inner tube, run the inner tube inside the outer tube, make as many inner tube connections as necessary, connect the inner tube plug on top, run the swivel assembly which is in contacts with the outer tube.
5. Run-in-hole without rotation and mud circulation (by making drill pipe connections). If there are any tight spots/interval, rotate the string to ream out.
6. About 90 ft (27 m) before reaching the bottom (off-bottom), start circulating, still without rotation.
7. When the formation was tagged, just pull up a few inches off-bottom while keeping the circulation (to clean the inner barrel and get ready to drop the ball).
8. Drop the ball (to change the mud path from inside the inner tube to the annulus between the inner and the outer tubes). Become ready for core-drilling.

   – Dropping the ball could be done from the surface or by pressure activation of the drop-ball sub above the inner tube.
   – Remove the Kelly or the top-drive prior to dropping the ball from the surface.
   – On average, allocate 1 minute for each 300 m for the ball to fall down.

---

[3]*Tugger line* is a wire rope used in the drilling rig for lifting light loads.

- When the ball is reaching the Pressure relief Plug (to seat there), decrease the circulation rate (Stroke Per Minute (SPM), or Gallon Per Minute (GPM)) so that the ball can seat properly.

9. For the first 20 cm, turn on the depth recorder, apply slow rotation (30–40 Revolutions Per Minute (RPM)) and low Weight-On-Bit (WOB) (1000 Ibs) on bottom and then, gradually increase.

10. Try to find optimal coring parameters (for more information refer to Sects. 9.2.4 and 9.2.5).

11. If the core barrel is short (only one joint for the Kelly system and one stand for the top-drive system), after finishing the coring, pull the string off-bottom (apply *overpull*[4]) until the core breaks (Weight-On-Bit (WOB) indicator shows a reduction in weight which shows the core has been broken and taken by the core catcher spring).

12. Raise the BHA a few meters off-bottom and come back to near-bottom. If no obstruction is observed and the weight has dropped, the core is in the inner tube. Otherwise, Pull-Out-Of-Hole (POOH).

13. If the core barrel length is longer (greater than one joint/ $\sim$ 30 ft using A) the Kelly system or greater than one stand/ $\sim$ 90 ft using B) the top-drive system):
A: Using the Kelly system, for long-core barrels, after coring one joint ($\sim$ 30 ft):

A-1. First, it is needed to make a drill pipe connection. Thus, we need to raise the Kelly so that the drill string (collar) is exposed and we can set the slips and clamps. Next, the connection is made while the core bit and BHA is off-bottom, which also causes the $\sim$ 30 ft core to break (undesirably increasing the possibility of core jam).
A-2. After making the connection, circulate the mud (to clean the well off the cuttings, or junk), and then resume coring.

B: Using the top-drive system:
We can continue coring nonstop until three joints ($\sim$ 30 ft) of core is taken. If the core barrel is longer than three joints, we need to first make a connection. Unlike the Kelly system, there is no need for making the core bit off-bottom and then make the connection. After the connection, we can continue coring.

14. At the end of coring, break the core (just like stage 11), and POOH with safe tripping rate/speed (refer Chap. 8, Sect. 8.3).
Note: It is advisable to set the slips gently to prevent core mechanical damage during POOH.

---

[4]For overpull, apply 5,000–35,000 Ibs depending on the core size.

5.4   Challenges in Coring Operations

As it was discussed in Sect. 2.3.2, there are six coring KPIs for evaluation of the success of a coring job. Five out of these six KPIs are technical: the coring safety, reliability, efficiency, recovery, core quality. The six KPI is the cost. Coring operations face some challenges which can lower the KPIs and thus question the success of the operations. There are several technical challenges that may occur to any coring job including the conventional ones. Table 5.1 lists some main technical challenges during coring operations, which include long coring, unstable cores (from shales, unconsolidated sand, faulted/fractured formations), wellbore instability, vibration, improper weight transfer to the bit in highly deviated wellbores, extremely low mud circulation rate, significant mud invasion, existence of bottom-hole fill prior to coring, etc. (inferred from Storms et al. 1991; Skopec and McLeod 1997; Whitebay et al. 1997; Silva et al. 1999; Hettema et al. 2002; Briner et al. 2010; Guarisco et al. 2011; Gay 2014; Hegazy et al. 2014; Mukherjee et al. 2015; Keith et al. 2016).

The potential consequences of not overcoming the aforementioned challenges are obtaining low coring KPIs: (a) mainly jamming/stuck of the sample which causes unprecedented coring termination, core damage or fractures and (b) core damage due to invasion. Jamming occurs during the core entry into the inner tube due to excessive friction between the rock samples and the inner tube. The friction becomes excessive particularly in case of unconsolidated rocks, already fractured rocks, when the weight transfer on the bit is not appropriate in highly slanted wells, or when the mud hydraulics is improper and cannot make the bit cutters clear off the cuttings during cutting. All these consequences signify low technical KPIs.

To overcome the mentioned challenges and reach a successful coring operation, there are some recommendations (inferred from *the same references*, and summarized in Table 5.1). For extremely long-core barrels (e.g., longer than 90 ft), it is recommended to use antijamming or telescoping core barrels (refer to Table 4.1) while using PDC core bits,[5] and properly design the BHA. However, it is recommended to keep the core barrel shorter than 120 ft particularly for troublesome formations such as in shales or fractured formations. For the challenge with unconsolidated or unstable cores, the design of inner tubes and using triple tube systems (i.e., utilization of an aluminum inner liner which has lower friction, inside the inner tube) and full-closure systems (look at Chap. 9) can enhance the coring efficiency and recovery. For the challenge of wellbore instability, it is recommended to create a geomechanical model of the core-drilling operation in advance to prevent any induced fractures in the formation prior to be cored. The modeling results in appropriate operating parameters such as WOB and RPM to be applied. In addition, the appropriate selection of the coring fluid matters for this issue. For the vibration issue, the BHA design and optimized operating parameters such as WOB and RPM

---

[5]PDC bits have longer life and thus contribute to reducing the number of round trips particularly for long coring operations with minimum core barrel length of 90 or 120 ft.

Table 5.1 Some technical coring challenges, consequences, and recommendations

	Challenge	Potential consequence(s)	Recommendation(s)
1.	Extremely long-core barrel	Possibility of jamming (particularly in fractured, shaly, and faulted formations), possibility of core damage	Use PDC bits if the formation is hard, use antijamming system, BHA design, keep the core barrel length shorter than 120 ft (in troublesome formations)
2.	Unconsolidated unstable cores	Causes severe core damage, or jamming	Use triple tubes (refer to Table 4.2), full-closure systems (refer to Chap. 9)
3.	Wellbore instability	Causes inefficient coring, core jamming, low core recovery and quality	Model the coring geomechanics (particularly in fractured formations), optimize operating parameters, apply efficient coring fluids with wellbore strengthening materials
4.	Vibration	Bit whirl and vibration causes coring-induced damage, increase jamming/damage possibility, etc.	BHA design, keep RPM/WOB low enough while core-drilling, construct geomechanical model for the vibration
5.	Improper weight transfer on bit	Due to stabilizers hanging on the wellbore, particularly in deep and inclined hole coring, core jamming, fracture, or core damage may occur	BHA and bit design, keep the inclination low enough in deep jobs
6.	Extremely low circulation rate	Possibility of getting off-bottom while core-drilling which leads to jamming (extremely low rate) Possibility of washing the core (extremely high rate)	Model the hydraulics prior to the operation, during operation, keep on-bottom (even during connections)
7.	Significant mud invasion	Invaded/damaged core	Use low-invasion or invasion–mitigation systems (refer to Chap. 7)
8.	Bottom-hole fill	Causes core jamming in the core bit, core catcher, or inner tube	Hole cleaning before POOH of the drill string, proper circulation through the inner tube prior to coring commence (for open-ended inner tubes), tag total depth to identify possible fill[a]
9.	Human error	Causes inefficient coring and low core recovery and quality	Provide efficient training (inferred from Lee et al. 2013)

[a]As part of a standard coring procedure, it is required to tag the bottom depth (on top of a possible downhole fill/junk) no matter if an open inner tube or a closed inner tube system is used

are significant to prevent excessive vibration. For the challenge of improper weight transfer on the core bit, it is important to properly design the BHA and the bit and keep the inclination angle low enough for deep coring. For the challenge with the

Fig. 5.4 A bit blade as the junk riding on top of the core sample (photo published courtesy of Baker Hughes GE)

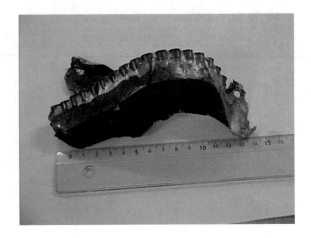

mud circulation rate, it is crucial to optimize the rate which can let sufficient cuttings removal to the surface. For the bottom-hole fill, it is recommended to conduct a good hole conditioning and cleaning before pulling out of hole of the drilling string and for open-ended tubes after running in hole of the coring assembly. Regardless of the use of open-ended or closed-ended/enclosed inner tubes, it is anyway part of a standard coring job to tag the bottom depth (high on top of any possible fill) to identify if there is any fill/debris at the bottom of the well before proceeding to coring. In Fig. 5.4, an example of a fill is shown, here a bit blade which has entered the inner tube on top of the 4-in core column. However, in this example no jam occurred.

Having discussed the technical challenges, the cost challenge is discussed. Coring cost is generally significant which can make the decision on coring uncertain or negative, as was already introduced in Sect. 2.2.2. In particular for conventional coring, the main challenge is the high cost incurred specifically by the long time required for tripping of the drilling and coring assemblies. This challenge causes poor KPI for cost. Unfortunately, there are no recommendations to seriously deal with this challenge of conventional coring. This challenge makes this method uneconomic and not an interesting option for coring for some cases, particularly deep cores.

References

Briner, A.P., A.H. Azzouni, R. Chitnis, and V. Vyas. 2010. *Sweet Success in Sour Coring*, SPE 128007 MS. Presented at the North Africa Technical Conference and exhibition, Cairo, Egypt, February 14–17.

Gay, M. 2014. *A Coring Matrix for Success*. Presented at the International Symposium of the Society of Core Analysts, Avignon, France, September 8–11.

Guarisco, P., J. Meyer, R. Mathur, I. Thomson, J. Robichaux, C. Young, and E. Luna. 2011. *Maximizing Core Recovery in Lower Tertiary Through Drilling Optimization Service and*

Intelligent Core Bit Design, SPE/IADC 140070. Presented at the SPE/IADC Drilling Conference, Amsterdam, the Netherlands, March 1–3.

Hegazy, G.M., A.S. Ragab, S.A. Shedid, M.N. Aftab, M. Ali, M.M. Al Riyami, H.W. Ibrahim, and M.F. Ibrahim. 2014. *Innovative and Cost-Effective Coring Technique "Extended Coring" for Long Intervals of Multiple Zones with World Record - Case Histories from the UAE*, SPE 171852 MS. Presented at the Abu Dhabi International Petroleum Exhibition and Conference, November 10–13.

Hettema, M.H.H., T.H. Hanssen, and B.L. Jones. 2002. *Minimizing Coring-Induced Damage in Consolidated Rock*, SPE-78156-MS. Presented at the SPE/ISRM Rock Mechanics Conference, Irving, Texas, October 20–23.

Keith, C.I., A. Safari, K.L. Aik, M. Thanasekaran, and M. Farouk. 2016. *Coring Parameter optimization-The Secret to Long Cores*. Presented at the OTC Conference, Kuala Lumpur, Malaysia, March 22–25.

Lee, R.K., P.M. Strike, C.D. Rengel, and D.B. Sutto. 2013. *Harnessing Multiple Learning Styles for Training Diverse Field Personnel in Conventional Coring Operations*, IPTC 16654. Presented at International Petroleum Technology Conference, Beijing, China, March 26–28.

Mukherjee, P., J. Peres, B.S. Al-Matar, P. Kumar, P.K. Choudhary, W.K. Al-Khamees, and M. Stockwell. 2015. *Piloting Wireline Coring Technology in Challenging Unconsolidated Lower Fars Heavy Oil Reservoir, Kuwait*. Presented at the SPE Kuwait Oil and Gas Show and Conference, Mishref, Kuwait, October 11–14.

Silva, A.J., W.R. Bryant, A.G. Young, P. Schultheiss, W.A. Dunlap, G. Sykora, D. Bean, and C. Honganen. 1999. *Long Coring in Deep Water for Seabed Research, Geohazard Studies and Geotechnical Investigations*. Presented at the Offshore Technology Conference, Houston, Texas, May 3–6.

Skopec, R.A., and G. McLeod. 1997. Recent Advances in Coring Technology: New Techniques to Enhance Reservoir Evaluation and Improve Coring Economics. Published in the *Journal of Canadian Petroleum Technology*, PETSOC 97-11-02.

Storms, M.A., S.P. Howard, D.H. Reudelhuber, G.L. Holloway, P.D. Rabinowitz, and B.W. Harding. 1991. *A Slimhole Coring System for Deep Oceans*, SPE 21907-MS. Presented at the SPE/IADC Drilling Conference, Amsterdam, the Netherlands, March 11–14.

Whitebay, L., J.K. Ringen, L.V. Puymbroek, L.M. Hall, and R.J. Evans. 1997. *Increasing Core Quality and Coring Performance Through the Use of Gel Coring and Telescoping Inner Barrels*, SPE 38687 MS. Presented at the SPE Annual Technical Conference and Exhibition, san Antonio, Texas, October 5–8.

Chapter 6
Wireline Continuous Coring

6.1 Introduction

As discussed in Chap. 5, conventional coring suffers from extremely long tripping times. The whole drilling string must be pulled out of the hole before the whole coring assembly (including inner and outer tube assembly) can be run in the hole. Similarly, following the core-drilling, the coring string must be conventionally pulled out before the drilling string can be run in the hole again to resume drilling. This causes a considerable waste of time during tripping. To address this issue, an alternative method of coring (i.e., wireline continuous coring) has been considered and applied in the petroleum industry (Walker and Millheim 1990; Randolf 1991; Deliac et al. 1991, Bencic et al. 1998; Warren et al. 1998).

It should be noted that wireline continuous coring is not a new topic in the petroleum industry, but rather it originates from the mining industry. It was translated and applied in the petroleum industry until the 1960s (refer to the patent information corresponding to the wireline continuous coring in Table 15.7). However, the old systems faced some practical problems which caused a halt in the operations until recently. In recent years, with the new systems, this method has become increasingly popular.

Therefore, in this chapter, following the description of the wireline coring including its schematics, its advantages and disadvantages are compared with other coring methods. Next, the enclosed ball and seat and the latching mechanism in the drilling and coring modes are covered. The chapter is continued by step-by-step practical procedure. Finally, a special wireline coring with navigation feature is discussed. As knowing about the wireline core barrels is crucial for understanding the contents of this chapter, Chap. 4 is considered as the prerequisite and reference for this chapter.

Depending on the lithology, depth, core size, etc., apply, e.g., 5000 to 35,000 Ibs overpull.

© Springer International Publishing AG, part of Springer Nature 2018
R. Ashena and G. Thonhauser, *Coring Methods and Systems*,
https://doi.org/10.1007/978-3-319-77733-7_6

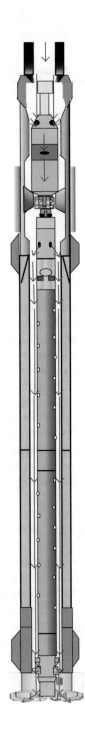

6.2 Description

As mentioned in Chap. 3, *Wireline Continuous Coring (WCC)* is a method of coring/drilling by which switching from drilling to coring and vice versa is performed via the wireline, without any conventional pipe trips required. To make this possible, both for drilling and coring modes, the same outer assembly, i.e., the Bottom Hole Assembly (BHA), and the core bit remain bottom-hole and just the right inner assembly is inserted for each mode. In other words, the slick line or wireline is used to trip any inner assemblies in and out of the hole. The inner drilling assembly includes a *drill bit insert/plug* to be fitted into the core bit to convert it into a full drilling bit (drilling mode). The inner coring assembly includes an inner tube assembly to be fitted into the outer assembly to receive the core sample as it is cut (coring mode).

Depending on the number of cores to be taken in a hole section, wireline continuous coring is composed of a sequence of several drilling and coring phases, which are switched/converted to each other via wireline trips. That is the reason for the name *continuous*.

Wireline continuous coring starts with the drilling mode. The schematic of drilling mode (the inner and outer assemblies) is shown in Fig. 6.1a, b. Figure 6.1a shows the mode before the locking grapple becomes hydraulically locked/latched. Figure 6.1b shows the mode after the locking grapple is locked/latched hydraulically and mechanically (i.e., collet fingers latching) above the inner tube. The hydraulic latching in drilling mode is done first by the hydraulic activation of the mandrel on the locking grapple, and second by the *lobe sub* which is hydraulically locked by the mud hydraulic pressure at the bottom of inner tubes. As another system of the inner tube assembly (in another design), a *drive latch* is used near the bottom of the inner drilling assembly to mechanically latch the inner drilling assembly to the outer drilling assembly (Fig. 6.2). The latching mechanism in the coring mode is explained in detail in Sect. 6.5.1.

In order to switch the drilling to the coring mode, the *drill rods* and the *drill insert/plug* (which was inserted inside the core bit) are extracted from the bottom hole assembly, BHA (which transforms the drilling bit into a core bit) and pulled up by the wireline and *overshot assembly*[1]; next, the inner tube/barrel assembly is tripped into the outer tube assembly by wireline. In Fig. 6.3, the schematic of the coring mode (including its inner and outer assemblies) are shown. Figure 6.3a shows the mode after the pressure head assembly has seated in its seat by the hydraulic mud pressure (in coring mode, only hydraulic latching is possible as

---

[1]Overshot is a latching device attached down the slick line/wireline at one end and at the other end is connected to the inner tube assembly or inner drilling assembly. It is used in order to unlatch or latch the inner drilling assembly for tripping out or in (at the end or beginning of the drilling mode) or to unlatch or latch the inner tube assembly (at the end or beginning of the coring mode).

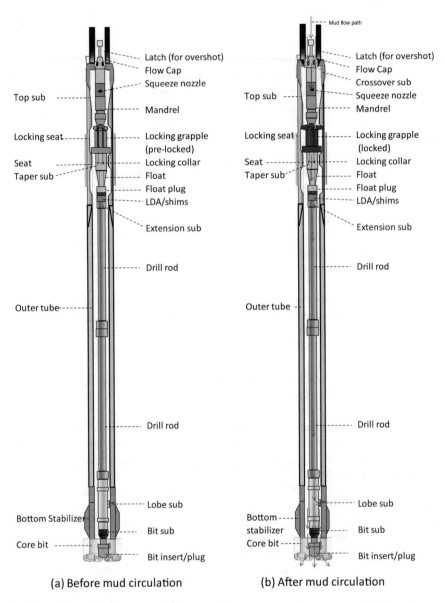

(a) Before mud circulation   (b) After mud circulation

Fig. 6.1 Drilling mode with the inner drilling assembly inside the outer tube assembly on the top and bottom, respectively, by mechanical and hydraulic latching assemblies, **a** prior to latching and drilling, **b** after latching and while drilling

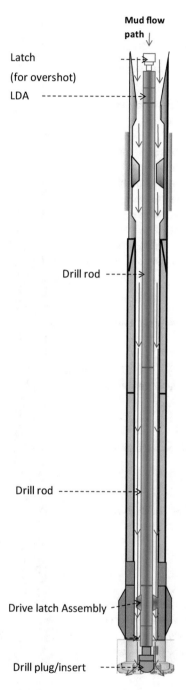

Fig. 6.2 Another type of inner drilling assembly in which the inner assembly is near its bottom at the drive latch assembly

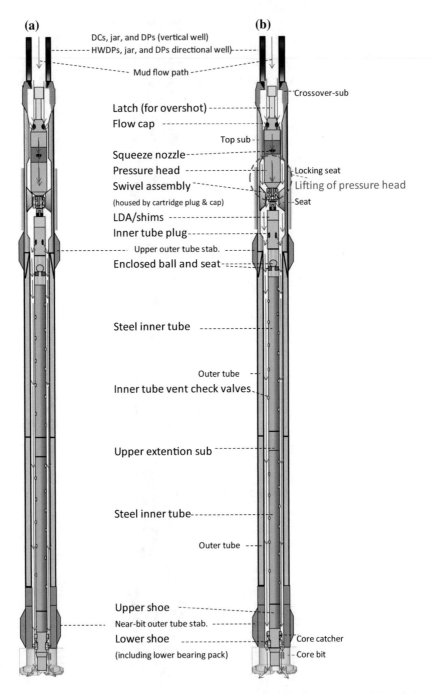

Fig. 6.3 Coring mode with the inner tube assembly inside the outer tube assembly **a** during coring/core-drilling, **b** just at the end of coring when the inner tube becomes full or core jamming occurs (which is signed by the lifting of the pressure head, the red circle)

explained in Sect. 6.5.2). Figure 6.3b shows the end of coring mode either when the inner tube is filled with the core column or when core jamming has occurred, which causes lifting of the inner tube.

In order to switch the coring to the drilling mode, first, the inner tube assembly (including the inner tube and the bearing assembly) is retrieved by the wireline and *overshot assembly*. Then, the retrievable inner drilling assembly (including the drill plug and drill rods) is conveyed in hole by wireline to drill ahead (drilling mode). It is also possible to run in an empty inner tube assembly (after recovering the sample inside at the surface) to cut additional cores (further coring). The cut-core is contained by the inner tube which is either out of steel, aluminum, etc., with a length of ~4.5 or 9 m (in petroleum applications), typically shorter than that of conventional coring.

It is noted that the combination of the core bit and the insert/plug bit should have enough stability to perform identically to a full drilling bit while drilling. Therefore, first, it is important that the insert plug is latched tightly to the core bit. Second, some drill rods are required to be placed on top of the drill insert to provide its required weight. Third, the drill rods must be attached to the outer tube so that the inner drilling assembly can turn with the same RPM as the outer tube assembly. Like conventional coring, prior to the deployment of the inner tube assembly into the hole, it is recommended to investigate the optimized values of the coring RPM and WOB for optimal performance.

6.3 Comparison with Other Methods

Comparing wireline continuous coring with other fundamental coring methods (refer to Chap. 3) shows several advantages and disadvantages of wireline coring as follows:

6.3.1 Advantages

The average wireline tripping rate is 300 ft/min ($\approx$1.52 m/s) whereas the quickest drill pipe possible tripping rate using the top-drive system is 1 minute per stand ($\approx$0.45 m/s). Thus, the wireline tripping rate is at least about 3.4 times that of the pipes. In wireline continuous coring, the trip time and thus the rig time is considerably mitigated owing to the greater wireline speed than the conventional pipe tripping rate. Therefore, less rig time would be required for wireline continuous coring method than the conventional one (Warren et al. 1998; Gelfgat 1994; Ali et al. 2014). This is analyzed as follows:

Basically, the total coring time consists of the following:

- Tripping time ($T_{Tripping}$)
- Core-drilling time ($T_{Core-Drilling}$)
- Handling time at the surface ($T_{Atsurface}$), which is the time required for the preparation and handling of the coring tools for each run.
- Non-Productive Time (*NPT*), e.g., which is the time due to waiting for the tools or tool failure.

Thus, total coring time (T_{Total}) is found as follows:

$$T_{Total} = T_{Tripping} + T_{Core-Drilling} + T_{Atsurface} + NPT. \qquad (6.1)$$

The time-related coring cost can also be found as follows:

$$C_{Coring-Time} = C_r \times T_{Total}, \qquad (6.2)$$

where C_r is the daily rig rate.

Using Eq. 6.1, the following relation presents a model for coring time:

$$T_{Total} = \frac{1}{24}\left(\frac{\Delta D}{L} \times \frac{TD}{V_t} + \frac{\Delta D}{ROP} + \frac{\Delta D}{L} \times T_s + NPT\right), \qquad (6.3)$$

where

T_{Total}: Total coring operations time [Day]

TD: Total depth after coring [m]

ΔD: Coring interval [m]

L: Core (barrel) length [m]

$\frac{\Delta D}{L}$: Number of required trips

V_t: Tripping rate [m/h]

T_s: handling time at the surface between two coring runs [h]

ROP: Rate of penetration [m/h]

NPT: Non-Productive Time [h].

Next, for a more general cost investigation, we use another cost term (coring cost per foot). This is found by (inferred from Samuel 2010) as follows:

$$C_{Total}[USD/ft] = \left(\frac{C_{Bit} + C_{service} + C_r \times (T_{Total})}{\Delta D} \right) \times \frac{1}{R_C},$$
(6.4)

where

C_{Total}: Total cost of the coring operations [USD/ft]

C_{Bit}: The cost of the core bit [USD]

$C_{service}$: Cost of the service-company's service [USD]

T_{Total}: Total coring operations time [Day]

ΔD: Coring interval [m]

R_{Coring}: The core recovery (at the surface).

Therefore, using the data given in Table 6.1 for a typical coring job in a greatly troublesome formation (e.g., fractured or unconsolidated), the total coring time (T_{Total}), the time-related coring cost ($C_{Coring-Time}$) and the total coring cost per foot have been evaluated, and compared for the conventional and wireline continuous coring methods. It shows that wireline continuous coring can contribute to reducing the coring cost from 8612 to 6967 USD/ft, particularly in troublesome formations. The excessive coring costs of this troublesome formation in this example is due to several trips incurred and using greatly short core barrels (which is necessary for such formations). It is noted that for a non-problematic formation, a standard coring job may cost as low as 200 USD/ft (~ 650 USD/m), or if a high-tech service is used, it can rise up to 700 USD/ft (~ 2300 USD/m).

Using the above example, the positive effect of the wireline tripping time is obvious on the whole coring time and cost. Therefore, as shown in Table 6.2, the wireline method is strongly recommended for coring in deep formations, for long core sections, for multiple zones, in exploration wells, when core points are unknown, in formations with high jamming probability, when logging is required following coring, and in out-of-gauge wellbores. In terms of costs, Sidewall coring is considered as a rival of wireline continuous method because it is run via wireline as well. However, the wireline continuous coring has several geological and reservoir advantages to this method as listed in Table 6.3.

Table 6.1 A typical comparison of time-related cost for continuous and conventional coring of a critical formation

	Wireline continuous coring	Conventional coring
Coring interval, ΔD (m/ft)	3657.6–3688 m (12,000–12,099.7 ft)	
Core recovery	90%	
Core (barrel) length, L (m)	6 m (20 ft)	10.6 m (30 ft)
Tripping rate, V_t (m/h)	7242 (120 m/min)	724.2 (12 m/min)
ROP (m/h)	5.03	20.12
Surface time per run T_s (h)	0.5	
NPT (h)	1	
Bit cost (USD)	15,000	
Coring service (USD)	100,000	
Daily rig rate (USD/Day)	150,000	
No. of coring runs	$\frac{\Delta D}{L} = \frac{30.4m}{6m} \approx 5$	$\frac{\Delta D}{L} = \frac{30.4m}{10.6m} \approx 3$
T_{Total}	$= 5 \times \frac{3688\,\text{m}}{7242\,\frac{\text{m}}{\text{h}}} + \frac{30.4}{5.03} + \frac{30.4}{6}\,0.5 + 1$ $= 12.1\,\text{h} = 0.5\,\text{Day}$	$= 3 \times \frac{3688\,\text{m}}{724.2\,\frac{\text{m}}{\text{h}}} + \frac{30.4}{20.12} + 3 \times 0.5 + 1$ $= 19.3\,\text{h} = 0.8\,\text{Day}$
$C_{Coring-Time}$ (USD)	75,000	120,000
C_{Total} (USD/ft)	6967	8612

Table 6.2 Drilling related advantages of wireline continuous coring to other methods, (inferred from Gelfgat 1994; Warren et al. 1998; Shinmoto et al. 2011, 2012; Ali et al. 2014)

	Good for:	Reason
1.	Deep formations	Much shorter trip time
2.	Long core sections	Enables recovering inner tubes with short trip times
3.	Multiple (separate) zones	Much shorter trip time and the possibility of easily switching from drilling to coring and back
4.	In exploration wells or when the core points are unknown	Enables proceeding immediately to coring
5.	In jam-prone formations (swelling shales, fractured, etc.)	Enables recovering short core sections (to prevent jamming) with several runs or if jamming occurs anyway, retrieve the jammed cores and quickly retry
6.	When logging after coring is required	When the reservoir/zone of interest is not known enough and logging may be needed, this may necessitate further drilling or drilling rat-holes following or preceding coring **Note:** A rat-hole is an extra hole drilled at the end of the well to allow the tools at the top of the logging string to reach and measure the deepest zone of interest
7.	In out-of-gauge holes	In such holes, even side-wall coring is ineffective because of the out-of-gauge characteristic of the wellbore wall; however for such case wireline coring is a good option less expensive than the conventional method

Table 6.3 Geological and reservoir advantages of wireline coring compared with side-wall coring

Good for:	Reason
Reservoir analysis	Enables obtaining relatively larger cores (than the sidewall method) with less damage due to mud invasion and mechanical effects, and greater core recovery
Sedimentological analysis	
Can potentially provide higher quality cores in probably critical formations	
For quick volumetric analysis and decision-making especially in unconventional reservoirs, e.g., Coal-Bed-Methane (CBM) or gas hydrates	Because of the greater representability of the cores (as of larger size and less damage compared with sidewall method), following a quick surfacing, rig site lab analysis can enable a volumetric analysis and decision-making

6.3.2   Disadvantages

Wireline continuous coring has several disadvantages compared with conventional coring as follows (Ashena et al. 2016a, b):

– *Smaller core size (diameter and length)*:
   Compared with conventional coring, traditional wireline bottom-hole coring could take only small core sizes because at the end of coring, the inner tube must pass through the bore of the drill pipes above the outer tube. The typical core size was 1–1½-in. which was less than the minimum required (1.81 in.) and thus had limited value for core analysis. This has been increased to 3 or 3½-in. using recent developments (Shafer 2013). Generally, in smaller cores, the amount and reliability of core analysis data are less than that with cores with larger sizes.
– *Specialty-drill pipes are required* (Farese et al. 2013a, b; Shafer 2013):
   In order to overcome the core size limitation, specialty-drill pipes with larger bores are required.
– *Special/larger bore jars are required* (Farese et al. 2013a, b):
   In conventional coring, if a jar is used, the only limitation for its ID is that the drop-ball should be able to pass. Thus, its minimum ID should be 1.25-in. However, in wireline coring, if a jar is used, the inner tube containing the core is retrieved via wireline through the jar. Thus, in order to prevent the core size limitation by conventional jars, special jars with larger bores are required.
– *Additional personnel is required*:
   Handling the wireline unit and the coring tools require the required experts.
– *No possibility of inner tube flushing prior to commencing coring*:
   With the currently available wireline tools, after tagging the bottom with the core barrel, it is not possible to flush the inner tube, prior to commencing to core the rock (refer to Sect. 1.4).
– *Slick line/wireline issues and torque* (Farese et al. 2013a; Ashena et al. 2016a, b):
   Handling issues of the slick line/wireline are possible twisting or torque (which may necessitate stopping occasionally) or even rupture during the tripping (particularly during POOH) of the inner assemblies to the surface.
– *Incompatibility with some systems*: Wireline continuous coring cannot be used with the anti-jamming system (Sect. 9.3) or the full-closure system (Sect. 9.4).

6.4   Enclosed Ball and Seat

In the currently available wireline continuous coring, there is no ball to be dropped from the surface or by hydraulic activation. Instead, there is an enclosed ball already in place in the inner tube assembly prior to coring. This installation prevents the mud from circulating through the inner tube before coring and can potentially cause any downhole junk to fill the bottom of the tube prior to coring; therefore, at the time of coring it may cause jamming. Recently, some measures have been initiated to modify the wireline core barrel such that a drop-ball sub is accommodated in lieu of an enclosed ball sub.

6.5   Latching Mechanism

The latching mechanism is required to keep the inner drilling assembly in place in the drilling mode and similarly to keep the inner coring assembly in place during the coring mode. The mechanisms differ in the drilling from coring modes (for better visualization, refer to Figs. 6.1 and 6.3). The mechanisms differ depending on the company design (Warren et al. 1998): Figs. 6.1 and 6.2 illustrate two different latching mechanisms for the drilling mode.

6.5.1   Drilling Mode

The components of the latching assembly in the drilling mode consist of the *pressure head assembly, squeeze nozzle/nozzle-half, mandrel, locking grapple, top sub locking seat*, and also *lobe sub* (find them in Fig. 6.1).

Wireline continuous coring initially starts with the drilling mode. To do this, the inner drilling assembly is run in the hole via slick line/wireline. When the latching assembly reaches the proper position (the bottom of the *locking grapple* reaches the top of the locking seat), the mud flow is inevitably diverted and directed through the nozzle-half due to the mechanical seal established. As the coring mud is passing through the nozzle-half with high pressure, a hydraulic thrust load is exerted on the mandrel making it move forward forcefully to the locking grapple. Therefore, the fingers of the locking grapple are forcefully opened out horizontally and locked in place in the locking sub. As a result, the inner drilling assembly is latched mechanically and hydraulically from the top. As the second latching mechanism or back-up locking system, the *lobe sub* is used beneath the inner tube assembly. The lobe sub is hydraulically latched to the outer tube assembly to keep the inner drilling assembly and drill insert/plug in the center of core bit, which makes the insert and the drill rods rotate along with the outer tube.

In another system design of the inner tube assembly, a *drive latch* is used near the bottom of the inner drilling assembly (inside the near-bit stabilizer from outside) in order to latch the inner drilling assembly to the outer drilling assembly so that the drill insert and the drill rods can rotate in unison with the outer tube (Fig. 6.2).

6.5.2   Coring Mode

Latching problems in the coring mode may inhibit the coring tool from obtaining a core sample and therefore cause waste of the rig time and loss of information. The latching mechanism may fail to latch the inner barrel in position prior to taking a core sample and thus the core may never enter the inner tube. Such a failure may not be readily distinguished from the surface. While the most common latching failure is that of the inner barrel failing to latch to the outer barrel, it is also possible to have an unlatching failure where the coring tool fails to be unlatched after the surface operator believes the inner core sample has been taken. Such a failure results in the need to pull the drill string with the attendant cost in time.

For the preceding reasons, it is a popular option to use only hydraulic (no mechanical) latching of the inner tube using the *pressure head* and the *locking seat* while core-drilling, as shown in Fig. 6.3. As the coring/drilling mud is passing through the *nozzle-half* with high pressure, a hydraulic thrust load is exerted on the *pressure head* to move it forward forcefully to be hydraulically seated or locked on the seat. However, as soon as the mud circulation is stopped, the thrust load on the pressure head is removed, the hydraulic latch is removed, and the inner coring assembly is ready to be retrieved.

6.6   Procedures

In the wireline method, there are two procedures corresponding to the coring and the drilling modes as follows:

6.6.1   Drilling Mode

When it is intended to core using wireline continuous coring system, we start with the drilling mode. The schematic of the drilling mode assembly (including its inner and outer tube components), for pre-latching and post-latching, has been shown in Fig. 6.1. The procedure for the drilling mode is explained step-by-step as follows:

1. Having ensured that the coring and core handling-related personnel have passed the training courses, arrange a pre-job meeting with them.
2. Rig-up the wireline.
3. Check the drilling BHA (i.e., *the outer tube assembly* and *the inner drilling assembly: drill bit plug/insert, drill rods, locking grapple, float, shims/LDA*, etc.) carefully as a pre-job requirement.
4. Pick up the first (bottom) outer tube joint (from the mouse hole) using the *lift bail* and *elevator* and connect it to the stabilizer, lower and tighten it into the PDC bit (using the *bit breaker*) and lower in the hole. Then, set the slips and clamps.

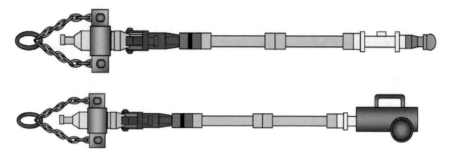

Fig. 6.4 Application of wheeled cart to the bottom of an inner drilling assembly for protection of the drill insert while raising it from the catwalk (through the V-door) to the rig floor

Note: It is noted that this PDC core bit has the possibility of receiving the drill pit plug/insert to be transformed into a full drilling bit.

5. Make as many outer tube connections as necessary, set the slips and clamps.
6. Pick up the drill rods from the ground to the rig floor through the catwalk using the *tugger/winch line*. Then, thread and tighten the drill rods on the drill bit plug/insert and the bit sub, which makes up the inner drilling assembly.
   Alternatively, lift the drill rods with their joints already connected to the ground, using the *wheeled cart* through the *catwalk* and *V-door* to the rig floor as a quicker method (Fig. 6.4).
7. Run the inner drilling assembly, i.e., drill rods and drill bit plug, through the outer tubes using the wireline and overshot, without any mud circulation.
8. When the inner drilling assembly reaches the proper position with respect to the outer tube assembly, start the mud circulation in order to activate the *locking grapple* to be locked into the *locking seat*. This mechanism makes the *drill bit plug* fixed in the core bit inner opening.
   Note: The BHA (the outer tube joints including stabilizers and the drill collars or heavy weight drill pipes, HWDP) will provide the necessary WOB on the core bit. The drill rods will provide the necessary WOB on the *drill plug*.
9. Release the wireline and overshot by shearing the *brass pins* (using the wireline jars).
10. Connect the drill pipes on top of the outer tubes to make up the outer tube assembly. Make connections and run-in-hole to reach one stand off-bottom from the Total Depth (TD) without rotation and mud circulation. In case there are any tight spots/intervals, rotate and ream during the trip.
    Note: Prior to drilling, always, the outer tube assembly must be one stand off-bottom.

11. Lower the outer tube assembly (to which the drill rods and drill bit plug are also attached) to the Total Depth (TD) while keeping the mud circulation.
12. Start the rotation and weight on bit (drilling). Optimize the drilling parameters (WOB, RPM, GPM, etc.).
13. Continue drilling until the coring point is reached.

6.6.2   Coring Mode

In wireline continuous coring, the drilling assembly is run first and then for coring, we switch to the coring mode (by replacing the inner drilling assembly by the inner coring/tube assembly). However, in critical situations (such as when core bit becomes dull), first, it is necessary to pull the drilling assembly out of the hole. Then, a new drilling mode assembly (including the inner drilling assembly) is run in the hole for safety reasons. This is because, unlike the coring assembly, the drilling assembly has a float valve which prevents entrance of any possible flow/kick from the formation inside the outer assembly while running and thus it provides some safety/well control. Next, following the replacement of the inner drilling assembly by the inner coring/tube assembly, coring is started. In the following, the detailed procedure for the coring mode (including the switching operations) is given as follows:

1. When the coring point is reached, stop the mud circulation so that the locking grapple can be unlatched. This is because stopping the circulation removes the hydraulic thrust load on the *mandrel*, which lets the locking grapple collet fingers become vertical by applying an upward force using the wireline.
2. If coring is to follow the previous drilling, first, pick up the drill string and the outer assembly for one stand off-bottom.
   2′ If the there is no drill string initially in the hole (e.g., if we want to core immediately after running and cementing the casing, or when there was already a necessary POOH due to the dull core bit, etc.), first, a new drilling outer assembly must be run in place only for safety reasons.
3. Once we reach the core point (whether it is at bottom of hole or after further drilling), the inner drilling assembly must be pulled out of the hole using the slick line/wireline and overshot:
   3a. Stop the mud circulation.
   3b. Using steel pin *overshot*, pick up the *wireline running gear*.
   3c. RIH with the wireline *running gear*, latch the running gear into the *rope socket*, *R-mandrel* or *tri-latch*, pick up and POOH to retrieve the mandrel, and drill rods back to the surface.

3d. Make up the *brass overshot* (in lieu of the steel one) to the bottom of the wireline.

Note: Brass overshots (with brass pin) are used for running the coring or drilling assemblies into the hole. In contrast, *steel overshots* (along with wireline running gear) are used for unlatching and pulling out of the drilling or coring assemblies.

4. Pick up the inner tube assembly (through the catwalk) using the brass overshot and wireline or alternatively, move the outer tube assembly using the wheeled cart through the *catwalk* to the floor as a quicker method.

5. RIH the inner tube (with the wireline and overshot) and latch in the inner tube assembly to the outer tube assembly.
   Note: Some coring companies[2] simply pump inner tubes downhole (in analogy with mining companies' strategy). This is called *pump-in systems*. Start mud circulation and pump the inner tube or coring assembly until the inner tube assembly seats in the *seat* of the outer tube assembly.

6. Shear the brass pin of the overshot to recover the wireline and overshot back to the surface.

7. POOH the wireline and overshot.

8. Lower the outer tube assembly to tag the formation, just raise the drill string a few inches off-bottom while keeping the circulation (to clean the inner barrel and get ready to drop the ball).

9. Drop the ball (to change the mud flow path from inside inner tube to the annulus between the inner tube and outer tubes and get ready for coring).

   9a. Remove the Kelly or top-drive prior to dropping the ball from the surface.
   9b. Roughly, allocate about 1 minute for each 250–300 m, for the ball to fall down.
   9c. When the ball is reaching the *pressure relief plug* to seat, decrease the circulation rate so that the ball can seat properly.

   Note: Dropping the ball is usually performed from the surface. In cases like motor coring, or generally for quicker operation, *drop-ball sub* can be used by which the ball is located just above the inner tube and is hydraulically activated by the hydraulic mud pressure surge.

10. Commence coring (also called core-drilling).

    10a. For the first 20 cm, turn on the depth recorder, apply slow rotation (30–40 RPM) and low WOB (1000 Ibs) on the bottom and then gradually increase.
    10b. Find and apply optimal coring parameters: WOB and RPM (refer to Sects. 9.2.4 and 9.2.5).

---

[2]Example NOV

11. If the core barrel is short (only one joint for the Kelly system and one stand for the top-drive system), after finishing of the coring, apply *overpull* (extra tension on the outer tube assembly) to pull the string off-bottom and break the core (the WOB indicator shows a weight-reduction which indicates the core has been broken and taken by the core catcher).

12. Raise the BHA a few meters off-bottom and come back to near the bottom. If no obstruction is observed and the weight dropped, we can ensure that the core is already in the inner tube. Otherwise, the core has not been taken and we need to POOH.

13. If the core barrel length is longer (greater than one joint/ ~ 30 ft using (A) the Kelly system or greater than one stand/ ~ 90 ft using (B) the top-drive system):

   A: *Using the kelly system, for long core barrels, after coring one joint (~ 30 ft):*

   A-1. First, it is needed to make a drill pipe connection. Thus, we need to raise the Kelly so that the drill string (collar) is exposed and we can set the slips and clamps. Next, the connection is made while the core bit and BHA is off-bottom, which also causes the ~ 30 ft core to break (undesirably increasing the possibility of core jam).

   A-2. After making the connection, circulate the mud (to clean the well off the cuttings, or junk), and then resume coring.

   B: *Using the top-drive system:*
   We can continue coring non-stop until three joints (~ 30 ft) of core is taken. If the core barrel is longer than three joints, we need to first make a connection. Unlike the Kelly system, there is no need for making the core bit off-bottom and then make the connection. After the connection, we can continue coring.
   Note: In general, when coring is resumed following the connection, the possibility of core jamming is high (because of the effect of stopping the rotation, circulation, and weight which is common between the Kelly and top-drive system, and mainly due to getting off-bottom with the Kelly system).

14. Finally, at the end of coring, apply *overpull* to break the core (just like stage 11), then POOH the inner tube assembly using the wireline and overshot, with safe tripping rate (Refer to Chap. 8, Sect. 8.3.1).

15. If it is intended to resume drilling after coring:

   15a. Pick up the string for one stand (i.e., one stand off-bottom) and run the inner drilling assembly again.

   15b. Remove the inner tube assembly from the outer tube assembly.

   15c. Pick up the inner drilling assembly through the catwalk using the tugger/ winch line.

15d. RIH the inner tube assembly and latch it into the outer tube assembly.
15e. Start drilling.
   Note: In general, before proceeding to either drilling or coring mode, the
   drilling or coring assembly must be placed off-bottom for one stand.

6.7   Navigated Wireline Coring

Wireline continuous coring is conventionally conducted without navigation.
However, at a time, a navigation feature was developed for it by including a near-bit
gamma-ray tool and directional sensors (inclination angle and azimuthal measurement) as shown in Fig. 6.5. The data could be transmitted to the surface via the

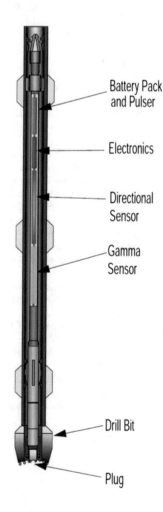

Fig. 6.5 Navigated wireline continuous coring with gamma-ray and directional survey in the drilling mode (published courtesy of Baker Hughes GE)

Battery Pack and Pulser

Electronics

Directional Sensor

Gamma Sensor

Drill Bit

Plug

wireline in a real-time manner. Thus, using this feature, it was possible to depth-match and to determine the coring points exactly. However, in recent years, this dated feature has been almost revived and turned into the Logging-While-Coring (LWC) feature, which has already been successfully practiced in some cases (for more information, refer to Chap. 12).

References

Ali, M., G.H. Hegazy, M.N. Aftab, A.M. Negm, A.A. Syed, and A.H. Anis. 2014. *First Wireline and Elevated Pressure Coring in UAE—Saved 30% of Coring Time for Shallow Reservoirs & Delivered Realistic Fluids and Gas Saturations*, SPE 171866 MS. Presented at the Abu Dhabi International Petroleum Exhibition and Conference, November 10–13.

Ashena, R., W. Vortisch, M. Prohaska, and G. Thonhauser. 2016a. *Innovative Concepts in Wireline Continuous Coring*, SPE 180017-MS. Presented at the SPE Bergen One-Day Seminar, April 20, Norway.

Ashena, R., W. Vortisch, M. Prohaska, and G. Thonhauser. 2016b. *Innovative Concepts in Wireline Continuous Coring*, SPE 0816–0060-JPT. Published in *SPE Journal of Petroleum Technology*, August, 60–61.

Bencic, A., M. Prohaska, J.T.V. De Sousa, and K.K. Millheim. 1998. *Slimhole Drilling and Coring-A New Approach*, SPE 49261. Presented at the SPE Annual Technical Conference and Exhibition, New Orleans, Louisiana, September 27–30.

Deliac, E.P., J.P. Messines, and B.A. Thierree. 1991. *Mining Technique Finds Applications in Oil Exploration*. Published in the *Oil and Gas Journal*, May, 85.

Farese, T.M., A.K. Mohanna, H. Ahmed, I.A. Adebiyi, and A.A.F. Omar. 2013a. *Coring Optimization: Wireline Recovery Using Standard Drill Pipe*, SPE 166739. Presented at the Middle East Drilling Technology Conference and Exhibition, October 7–9.

Farese, T., H. Ahmed, and A. Mohanna. 2013b. *A New Standard in Wireline Coring: Recovering Larger Diameter Wireline Core Through Standard Drill Pipe and Custom Large Bore Jar*, SPE 163507. Presented at the SPE/IADC Drilling Conference and Exhibition, Amsterdam, The Netherlands, March 5–7.

Gelfgat, M. 1994. *Complete System for Continuous Coring With Retrievable Tools in Deep Water*, SPE 27521 MS. Presented at the SPE/IADC Drilling Conference, Dallas, Texas, February 15–18.

Randolf, S.B., and A.P. Jourdan. 1991. *Slim-hole Continuous Coring and Drilling in Tertiary Sediments*, SPE 21906. Presented at the SPE/IADC Drilling Conference, Amsterdam, The Netherlands, March 11–14.

Samuel, R. 2010. *Formulas and Calculations for Drilling Operations*. 1st edn. Wiley-Scrivener. ISBN-13:978–0470625996.

Shafer, J. 2013. *Recent Advances in Core Analysis,* Published in the *Journal of Petrophysics*, 54 (6): 554–579.

Shinmoto, Y., E. Miyazaki, K. Wada, and M. Yamao. 2011. *Development of a Continuous Directional Coring System for Deep-Sea Drilling*, SPE 140913-PA. Presented at the EuroPEC/ EAGE Annual Conference and Exhibition, Vienna, Austria, May 23–26..

Shinmoto, Y., E. Miyazaki, K. Wada, and M. Yamao. 2012. *Development of a Continuous Directional Coring System for Deep-Sea Drilling*, SPE 140913-PA. Published in the *SPE Journal of Drilling and Completion*, March.

Walker, S.H., and K.K. Millheim. 1990. *An Innovative Approach to Exploration Exploitation Drilling: The Slim-Hole High-Speed Drilling System*, SPE 19525-PA. Published in *Journal of Petroleum Technology*, September, 1184, Translated by AIME, 289.

Warren, T., J. Powers, D. Bode, E. Carre, and L. Smith. 1998. *Development of a Commercial Wireline Retrievable Coring System*, SPE 52993-PA. Published in the *SPE Journal of Reservoir Evaluation and Engineering*, December.

Chapter 7
Invasion–Mitigation Coring

7.1 Introduction

One of the basic coring objectives is obtaining core samples with least possible damage to ensure reliable core analysis results. One of the sources of damage is related to mud invasion/chemical, which is introduced as follows.

During coring, the mud pressure is greater than the formation pressure in an overbalanced condition. Therefore, the mud or its filtrate may enter the core sample and displaces some of the movables fluids in the pores. During the filtration with water-based muds, a layer of impermeable mud cake is formed over the outside of the core sample and also penetrates a little through the core surface until bridging and sealing process is over. Indeed, the filtrate volume inside the core is mostly due to the spurt loss[1] phenomenon while the core is cut from the formation, which depends on the mud properties, the solid particle size, and the formation pore size distribution. A core sample obtained in this way is called *invaded*. Mud filtration causes uncertainty in the core analysis results as it adversely alters the saturation data, in situ rock wettability, and also affects the critical reservoir engineering and petrophysical parameters (such as residual oil saturation S_{ro}, Archie saturation exponent (n) and the relative oil and gas permeability K_{ro} and K_{rg}). Therefore, a large discrepancy may be observed between the water/hydrocarbon saturations from the core data and well logs. Particularly in exploration/discovery wells, this causes a large error in the hydrocarbon volume estimations.

Therefore, it is important to mitigate the invasion to the core, which is discussed in this chapter. To achieve this, first, the areas where invasion occurs should be identified. Next, the core size should be selected large enough (diameters greater than 2-in.). Then, the application of invasion–mitigation coring systems must be placed in priority. Generally, three invasion–mitigation coring systems are available in the market, which are *low-invasion coring*, *gel coring*, and *sponge-coring*. These

---

[1]Spurt loss volume is the instantaneous mud filtrate volume entering the formation just prior to the formation of any mud cake around the rock.

© Springer International Publishing AG, part of Springer Nature 2018
R. Ashena and G. Thonhauser, *Coring Methods and Systems*,
https://doi.org/10.1007/978-3-319-77733-7_7

Fig. 7.1 The outer and the central parts of a typical 4-in. core obtained by conventional tools. For example, the central part with 2½-in. for a 4-in. core may be uninvaded

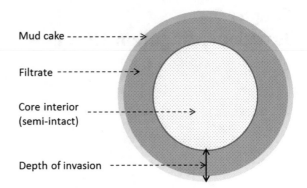

will be covered in this chapter. Although, invasion–mitigation systems reduce the value of invasion they do not eliminate it. Figure 7.1 shows a schematic of a recovered core sample using a filtrate mitigation system. Therefore, it is crucial to conduct core analysis experiments on the least uninvaded part of the sample, which is indeed the center of the core. This is considered its most valuable part for RCAL and particularly SCAL. As an example, in the 4-in. sample shown in Fig. 7.1, the semi-intact core-interior diameter can be 2½–3-in using a low-invasion system (inferred from Rathmell et al. 1999).

It is noted that all these methods discussed in this chapter remove only the invasion-related source of core damage, but the other source is mechanical which will be discussed in Chap. 8.

7.2 Invasion Areas and Factors

Filtrate invasion generally occurs during acquisition (coring) and also in case of careless handling and washing of the sample at the surface. During the acquisition, there are three areas at which mud invasion occurs to the core (as shown in Fig. 7.2):

– *Ahead the bit*:

The invasion ahead the bit occurs dynamically (as the mud is circulating) when the core sample is being penetrated by the core bit. It is significant with great overbalance pressure (which is the mud pressure minus the formation pore pressure) and when the coring rate of penetration (ROP) is much lower than the mud flow velocity into the core sample. Thus, optimized high enough coring ROP contributes to less invasion.

– *At the bit face and its throat (before the core enters the inner tube)*:

In overbalanced coring, the invasion at the bit face and throat occurs dynamically as the sample arrives at the bit face and its throat (just at the bottom of the

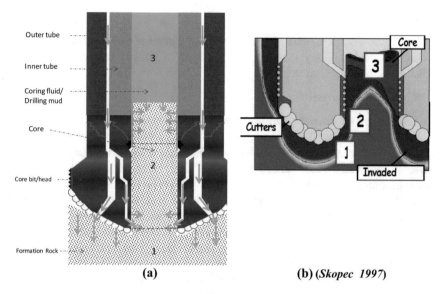

Fig. 7.2 The three principal areas where mud invasion occurs during coring, consisting of: (1) ahead the bit, (2) at the bit face and throat, (3) in the inner tube, shown in two views

inner tube). This invasion becomes more severe at low ROP; but it is significantly reduced when bridging solids and fluid-loss-controller materials are used in the mud.

– *In the inner tube (after the core has entered the inner tube)*:

In conventional coring, the mud is already in the inner tube when the core enters. In the inner tube, the invasion occurs as the mud pressure still exceeds the core pore pressure. This is considered as a static invasion as the mud is not circulating in the inner tube after the core has already entered the inner tube. After core drilling, this invasion ends when the inner barrel has been pulled out to a depth at which the mud pressure equals the core pore pressure. In addition to the overbalanced nature of this invasion, it can occur due to countercurrent imbibition.[2] The lower the underbalance pressure, the higher the ROP is or the quicker the POOH starts, the less the invasion would occur.

There are three factors affecting the amount of filtrate invasion, which are discussed as follows:

---

[2]Countercurrent imbibition is the process by which the wetting phase enters the rock (e.g., water penetrates into a water-wet rock) and the non-wetting phase (e.g., oil in a water-wet rock) escapes the rock in the opposite direction.

- *Formation properties*:

The formation properties affecting the mud filtrate invasion are permeability, wettability, and capillary pressure, and the type of fluid in the pore space (gas, volatile oil, heavy oil, or water).

- *Core bit/head properties*:

The amount of invasion also depends on the bit characteristics. Core bits may have a face-discharge efficiency which diverts or bypasses the mud flow from direct contact with the sample while its entry into the inner tube, which reduces the mud invasion at area-1 (i.e., ahead the bit). Next, core bits traditionally had Inside-Diameter (ID) gauge cutters/diamonds[3] to maintain the sample in-gauged. This caused the filter cake around the core to be removed and therefore caused extensive mud flushing and invasion, this feature should be revised. Next, the spacing between the core bit ID and the lower shoe affects the amount of the mud invasion at the bit throat (i.e., area-2). The greater this spacing, the greater the mud volume and invasion which can pass to invade the core sample on its way to enter the inner tube. The next characteristic is the length between the core bit throat and the bottom end of the lower shoe. The greater this length, the greater the exposure to the mud at the bit throat (i.e., area-2) and thus the greater the mud invasion.

Next, the bit properties may indirectly affect the mud invasion. Depending on the design of the bit and the formation properties, the operating properties such as WOB and RPM can be adjusted to optimize the ROP. The lower the ROP, the greater the invasion. Next, the mud circulation rate also partly depends on the core bit type and design. Coring with lower circulation rate keeps the (dynamic) bottom-hole Equivalent Circulating Density (ECD) and the circulating pressure lower (with fixed mud properties and Equivalent Mud Weight, EMW). Therefore, the lower bottom-hole pressure contributes to lower overbalance pressure and less invasion.

- *Mud properties*:

The amount of invasion also depends on the mud properties such as the mud type (water base muds, oil base muds, etc.), spurt loss (which is mud loss prior to mud cake creation), filtration rate, mud weight, and the concentration of the bridging materials inside the mud. The invasion typically matters for the water-based muds whereas in oil-based muds, the invasion is usually minimal and not a problem. Depending on the EMW, the circulation rate, and the evaluated ECD, the bottom-hole pressure and the differential pressure are found. The greater the spurt loss and the filtration rate are, the greater is the invasion. The greater the

---

[3]The gauge diamonds or cutters are generally used to protect the core bit diameter from becoming smaller. Previously, the inside-diameter gauge diamonds were traditionally used to help the ID of the bit maintain in-gauged (not to become smaller). However, they cause the mud cake around the core to be removed and thus more filtrate flushing would occur.

differential pressure is, the greater is the invasion. The greater the concentration of the bridging materials in the mud, the greater the invasion.

Based on the formation properties, the core bit and mud properties can be designed to mitigate the amount of invasion. This constitutes the basic of low-invasion coring (in the next section).

7.3   Low-Invasion Coring System

Low-invasion coring is considered the most fundamental invasion–mitigation system. A low-invasion coring system must have the following three features (inferred from Rathmell et al. 1994):

- Specially designed core bits/heads and extended inner tube shoes, which reduce the mud invasion
- Coring muds with special formulation (e.g., with low fluid loss, etc.)
- Optimized core drilling rate, such that the if possible coring ROP can be greater than the mud filtration rate.

Using this system, lower invasion occurs which keeps the central portion of the core uninvaded and thus proper for analysis (refer to Fig. 7.3). Therefore, rather reliable measurements of fluid saturations, wettability, relative permeability, and capillary pressure can be obtained from this portion. In better words, the validity and reliability of the results obtained from RCAL and SCAL are increased (inferred from Rathmell et al. 1999; Dennis 1999). Additionally, when a low-invasion coring system is combined with other invasion–mitigation methods such as gel or sponge

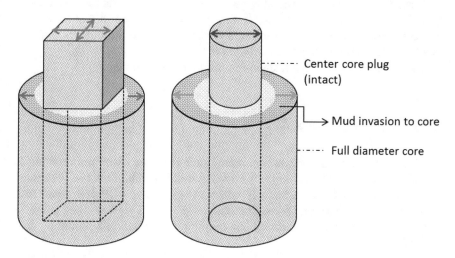

Fig. 7.3 The schematic of less-invaded core plug taken from a typical low-invasion coring system, which provides a central uninvaded portion for core analysis. This is useful for Routine Core Analysis, RCAL and Special Core Analysis, SCAL

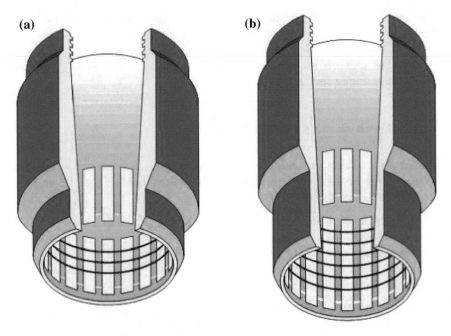

Fig. 7.4 a Conventional lower shoe and **b** extended lower shoe for low-invasion coring (published courtesy of Baker Hughes GE)

coring, more reliable data can be obtained. Currently, this system is commonly used by most systems such as pressure/in situ coring, oriented, etc., as one of their inevitable features. Nevertheless, this system requires an extra cost for the mud with special additives, the core bit, etc. It also requires an additional communication between all the involved parties such as the drilling company, the logistics, and etc.; otherwise, the system cannot properly lower the invasion.

The aforementioned three features are discussed as follows:

7.3.1 Optimized Core Bits

Low-invasion core bits/heads are optimized such that they lower the mud filtrate invasion, as shown schematically in Figs. 7.5 and 7.6. PDC core bits, compared with roller-cone bits, can potentially contribute to greater ROP and lower mud invasion particularly in hard formations and deep wells. Therefore, they are more commonly used in coring practices; however, their characteristics should be optimized as follows:

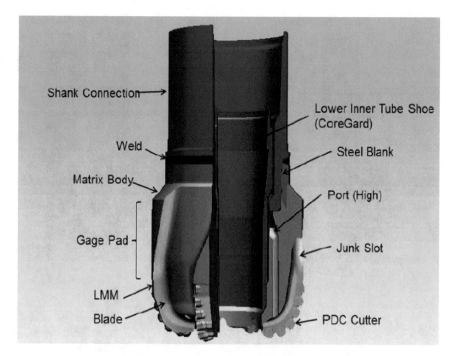

Fig. 7.5 A schematic of a low-invasion core bit and the pilot/lower shoe to show the minimal difference between the Inside Diameter (ID) of the bit and the Outside Diameter (OD) of the inner shoe (i.e., the lowest clearance for the mud to pass from the bit throat to the bit) (published courtesy of Baker Hughes GE)

1. *Parabolic bit profile*:

The flat profile and also long parabolic profile are not generally appropriate for coring because they are not the least aggressive and the resultant ROP is significantly low. Instead, parabolic bit profile is usually selected because first, it reduces the dynamic filtration area for invasion and second, it is more aggressive than the flat profile and can thus provide greater ROP, which in turn reduces the mud invasion.

2. *Aggressive cutter design*:

Low-invasion core bits should have a degree of aggressiveness of cutters by reducing the number of the PDC cutters, rather low *back-rake angle* (the angle of the cutter with respect to the vertical), and using large diameter PDC cutters. These features provide increased depth of cut and higher ROP and therefore lower invasion rate. It is noted that some limitations for these parameters should be regarded for hard formations; otherwise, extremely high aggressiveness can cause premature wear of the cutters.

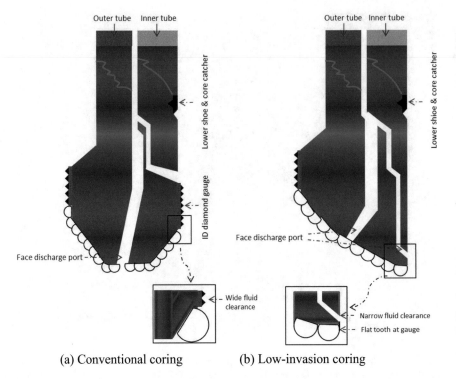

(a) Conventional coring   (b) Low-invasion coring

Fig. 7.6 Schematics comparing: **a** a typical conventional core bit and **b** a low-invasion core bit

3. *Minimal number of (OD) gauge cutters*:

Low-invasion bits have a minimal number of gauge cutters because it helps to reduce the contact time of gauge cutters on the core sample and increases the ROP. These result in lower mud invasion.

4. *Elimination of throat/ID cutters*:

ID-gauge cutters (e.g., made of diamonds) scratch the mud/filter cake around the core. Elimination of these cutters helps to reserve the mud cake, reduce the coring time, and increase the ROP. These signify lower mud invasion to the sample.

5. *Extended pilot/lower shoe*:

The lower shoe or the pilot (catcher) shoe is the sub which is attached to the core bit and contributes to breaking the bottom of the core at the time of over-pull of the coring assembly. When the lower shoe is extended (as shown in Fig. 7.4), it would be closer to the bit throat. In other words, the bottom end of the inner tube can be located nearer to the bit face and throat area to ensure that the core would enter the inner tube immediately after being cut. Thus, less invasion would occur to the core.

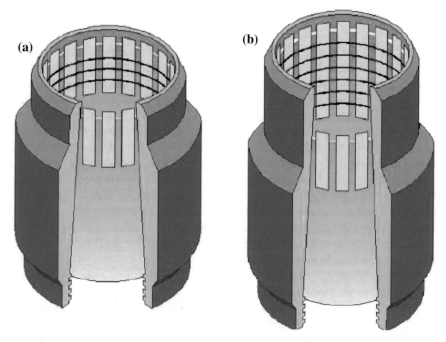

Fig. 7.7 a Conventional lower/pilot shoe, and **b** extended/low-invasion lower shoe (published courtesy of Baker Hughes GE)

6. *Face-discharge ports*:

Face-discharge ports are the conduits in the core bit for the mud to discharge and push the cuttings away from the bit. They must be directed away from the center of the bit in order to reduce the direct mud invasion to the sample as it is being cut.

7. *Minimal fluid clearance*:

The minimal clearance should exist between the core bit ID and the lower shoe OD. Thus, very limited mud volume can pass through the clearance to invade the core sample on its way to enter the inner tube.

8. *Low friction, anti-whirl bit*:

The core bit and BHA should be specially designed based on the formation rock properties such that the whirl and vibration can be minimized. This contributes to higher ROP and thus less mud invasion. In addition, they contribute to reduced induced fracturing of the core sample during the breakage, which means greater core quality (Fig. 7.7).

7.3.2 Optimized Mud Properties

The design of the coring mud formulation is greatly important for low-invasion coring systems. If the formulation is not appropriate, it is necessary to modify or even replace the drilling mud just prior to coring. Using specially designed coring muds (with special formulation), first, the coring mud density is kept lowest possible so that the overbalance pressure is kept the lowest. The mud spurt loss volume must be minimized by keeping the coring mud density and thus the overbalance pressure the lowest possible, while maintaining the plastic viscosity high enough. Plastic Viscosity (PV) represents the mechanical viscosity in the Bingham plastic model. In low density/solid concentration muds, it can be maintained high enough by adding polymers such as CMC or XC-polymer. Next, by controlling the mud rheological properties, the mud filtration rate is controlled such that a protective impermeable mud cake can be formed around the core sample. Usually, Calcium Carbonate ($CaCO_3$) with engineered particle size distribution is added to the mud to create an impermeable mud cake around the core column. The optimal particle size should range from 1.3 to 1.5 times the mean pore throat size so that they can properly bridge the pore openings.

Figure 7.8 shows typical macroscopic and microscopic plan-views to pinpoint that the depth of invasion of the mud filtrate is reduced using the low-invasion mud. With the preventive measures made against the invasion in this system, still some filtrate invasion may occur to the sample around the sample; however, its central portion remains almost uninvaded.

In practice, knowing the optimized coring mud properties during coring in a specified field still requires adequate research and investigation. For validation of the research results, it is sometimes necessary to use pilot tests to examine the performance of a designed mud. Therefore, it is important to quantify the invaded filtrate using the selected mud in order to investigate the effects of some factors such as the efficiency of the selected bridging solids, filtrate velocity versus estimated

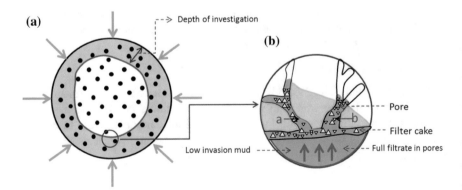

Fig. 7.8 The depth of invasion of the drilling mud in the cut core sample (plan view) **a** in macroscopic plan view and **b** the microscopic plan view

ROP, expected filtrate loss during the cutting, etc. This can be accomplished by adding tracers to the drilling mud before coring commences. Laboratory experiments can then detect the concentration of the invaded fluids inside the sample as they contain tracers. In water-based muds, the tracers used are chemical salts (such as NaCl, NaBr, NAI, KI, $NaNO_3$), stable isotopes (D_2O or heavy water), or radioactive isotopes such as tritium in case of having a license from Atomic Energy Commission AEC. The analytical methods of detection of the chemical salts, stable isotopes, and radioactive isotopes are respectively ion chromatograph, mass spectrometer, and scintillation counting spectrometer. In oil base muds, the tracers used are some specially selected hydrocarbons such as Iodonaphthalene, tritiated hexadecane (as an effective radioactive material), etc. For detecting Iodonaphthalene and tritiated hexadecane, respectively, a gas chromatograph equipped with an electron capture detector (GC/ECD) and scintillation counting spectrometer are used.

7.3.3 High ROP Relative to Invasion

In low-invasion coring, the coring ROP is kept as great as possible (but less than the upper allowable limit) so that it can be greater than the mud flow velocity, which prevents the invasion. If this is not technically possible (which is mostly the case), its difference from the mud flow velocity is reduced, which contributes to reducing the dynamic invasion to the core. It is noted that the coring ROP should be maintained lower than the upper allowable limit to prevent inducing fractures to the underlying formation prior to core drilling.

7.4 Gel Coring System

Although the low-invasion coring system contributes to effective minimization of the dynamic mud invasion to the core sample, there is still static invasion inside the inner tube and the wettability is altered (Whitebay et al. 1997; Skopec and McLeod 1997). Besides, during the surface handling including washing of the sample, the core cannot be protected against mud invasion. Gel coring is, in reality, an improvement to the low-invasion coring technique by gel-encapsulation of the sample in order to obtain uninvaded samples. It not only protects the sample during the cutting and tripping/retrieval, but also protects it during wellsite handling. Figure 7.9 compares the degree of invasion using low-invasion and gel coring. In low-invasion coring, the depth of invasion into the core is about 0.5–0.75-in., whereas a proper gel coring job can almost inhibit the invasion. This system is recommended particularly when the low-invasion system may not be properly implemented due to some problems such as the following:

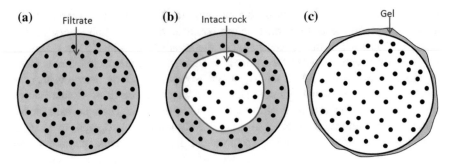

Fig. 7.9 Drilling mud invasion using three coring systems using **a** standard core head/bit, **b** low-invasion coring, and **c** gel coring with the low-invasion system

- Proper planning and communication for low-invasion coring have not been performed.
- The coring mud does not have ideal characteristics for low-invasion.
- The drilling conditions may not be adequately optimized or when just low ROP can be achieved due to the formation properties.
- Formation properties may cause some uncontrollable issues (particularly for exploration drilling).

7.4.1 Description

The gel coring system is considered a complementary invasion–mitigation system which uses polypropylene glycol based gels for encapsulating the core sample to mitigate the exposure or contact with the mud and thus reduces the invasion. The gel generally should have high viscosity, high molecular weight, zero spurt loss, and should be nonsensitive to temperature and pressure change, non-dissolvable in water, compatible with oil and mud, nontoxic, and environmentally friendly.

In this system, the inner barrel is filled with the gel before running-in-hole, and is held in place by a special piston. Thus, for gel coring, the inner tube must be modified to accommodate a simple *floating piston* (also called the *gel rabbit*) for gel distribution and encapsulation. As shown in Fig. 7.10, when the core is entering the inner tube, the corresponding valve at the bottom of the inner tube gets opened to allow the gel flow between the core and the inner tube, it encapsulates around the core sample to act as its protecting semi-permeable barrier (Skopec and McLeod 1997). After activation of the rabbit by the top of the core column, (1) the gel replaces the volume of mud in the area of the cutter to rock contact and therefore would partially prevent the dynamic invasion to the rock which is being cut, and (2) prevents the static invasion to the core inside the inner tube. As coring continues, the evacuation of the gel out of its place (to encapsulate) the sample

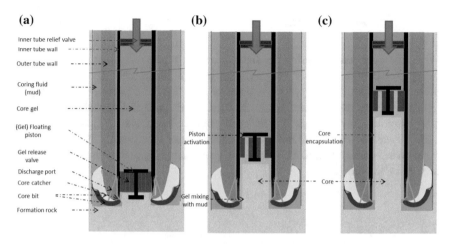

Fig. 7.10 The working mechanism of the gel coring system. **a** The inner tube *floating piston* or *gel rabbit* is closed before coring commences, **b** *Gel release valve* opens after being touched by the core and **c** Gel encapsulates and preserves the core. (Modified from Skopec and Mcleod 1997)

continues and the piston is being pushed up by the core. This invasion–mitigation is greatly essential particularly for highly permeable or gas condensate-bearing rocks, which are greatly susceptible to mud invasion. The core length can be variable, but field practice with 54 m was recorded.

During the core drilling, most of the original gel volume exits the barrel. During the retrieval/POOH, the gas expands out of the sample (due to pressure and temperature drop). This expels some of the gel between the inner tube wall and the sample out of the inner tube (through its check valves or the top). However, when the core sample is recovered at the surface, still a coat of gel remains around the core (e.g., about 5% of the original gel volume). Therefore, at the surface, it can preserve the sample and even mitigate possible invasion at the surface (in case of careless washing of the core) and can reduce its exposure to the air. These all reduce the pore fluid loss and wettability change. It is noted that just prior to lab preservation at the surface or conducting the core analysis, the gel must be taken off just like taking off the mud cake. This is because gels are not appropriate for long-term core preservation (e.g., for weeks or months).

In addition to mitigation of the mud invasion contributing to the core quality, gel coring has additional advantages. The gel lubricates the core sample and prevents jamming in jam-prone formations such as fractured ones, and thus increases the coring efficiency and core recovery. In unconsolidated rocks, it also enhances the mechanical strength and integrity of the core samples (Skopec and McLeod 1997).

7.4.2   Closed Inner Tube

When the core starts entering the inner tube, the rabbit will open the flow of gel being pressed along the core towards the bit and encapsulate the core while the pressure relief valve at the top will bleed of pressure if the pressure is building up inside the inner tube. Therefore, in this system, the ball and seat above the inner tube assembly have been replaced by a pressure relief valve. Therefore the gel coring system is run in the hole with a closed inner tube (unlike the conventional open-ended barrel); therefore, there is no possibility to circulate through the inner tube prior to coring.

7.5   Sponge Coring

Typical coring cannot provide accurate oil and water saturations including the initial or the residual oil saturations (representing the amount of oil in the reservoir) because of mud invasion and liberation of oil during tripping. Knowing accurate saturation values enables finding the fluid contacts (gas–oil contact and oil–water contact), the reservoir thickness, and the original oil in place. They are considered critical parameters for reservoir development strategies such as locating the optimal reservoir zones for perforation and fracturing. Therefore, sponge coring became available in 1981 with the objective of determining the in situ residual oil and water saturations in oil-bearing reservoirs by tightly surrounding the core and absorbing the liberated oil (inferred from Park and Devier 1983; Al-Housani et al. 2012). Additional advantages of sponge coring are mitigation of mud invasion and providing mechanical integrity of the unconsolidated cores as it is tightly surrounded by the sponge.

7.5.1   Description

In sponge coring, a tough porous polyurethane preferentially oil-wet sponge or foam is utilized as an outer sleeve in a disposable liner so that it can absorb and trap the oil expelled from the core material due to the pressure drop during tripping/retrieval. In order to convert a conventional core barrel into a sponge one, the following modifications must be made.

1. A disposable aluminum liner containing oil absorptive sponge is placed inside the inner tube. The maximum length of the inner tube used in this system is typically about 9 m, with ten aluminum sleeves of each 0.9 m.
2. The core size that can be obtained using sponge coring is smaller than with conventional methods. Therefore, a core bit with a smaller inner diameter should be selected. For example, using a conventional system with the dimensions of

7 7/8-in. * 4½-in., the ID of a core bit is 4½-in. whereas in a 8½-in. hole, in a sponge system with dimensions of 7 7/8-in. * 3½-in., the ID of the core bit is 3½ (i.e., one inch smaller).
3. A smaller and different core catcher is selected.
4. Similar to gel coring, the conventional ball and seat are replaced by an enclosed ball and seat or a pressure relief valve above the inner tube to let the excessive pressure of gas and water vent.

The sponge coring operations are similar to conventional/wireline continuous coring, except for some slight differences. The main difference is that the dropping of the ball or its activation is eliminated as the conventional ball and seat are not used in the system. The other differences correspond to the installation of the sponge liner in the inner tube and its handling after the retrieval. Immediately after the retrieval of the core barrel, each sponge liner is opened and wrapped in plastic, stored inside their PVC transportation tube, and transported to the lab. Recent research on the appropriate sponge properties has indicated that the sponge must be as follows:

– Capable of tolerating temperatures about 195 °F (90 °C) and high pressures.
– Chemically inert.
– Greatly porous to hold the expelled oil (>80% porosity).
– Greatly permeable to let the flow of oil inside (~2 D).
– Oil-wet to absorb the oil expelled during tripping.
– Flexible to be molded in the inner tube.
– (Forcefully) pre-saturated with brine using a vacuum pump in order to prevent the creation of a dry filter cake (e.g., of ~½-in. thickness) which can be a barrier against the oil flow from the core to the sponge.
– (Along with its aluminum holding liner) already-cut by *laser* or *plasma* for quick and easy core recovery at the surface.

As sponge coring is rather more costly than conventional coring; thus, its specific applications should be known. This system is strongly recommended for use across the transition zones to evaluate the saturation and identify the potential oil-bearing reservoir. Classifying this application based on the production phases, it is generally recommended:

– In the exploration phase to evaluate the saturation, fluid contacts, and finally the initial oil in place.
– In the secondary recovery programs to evaluate the effects of gas injection, water injection, etc.
– In the tertiary recovery programs to evaluate the effects of CO_2 injection, polymer flooding, steam injection, etc. (Figs. 7.11, 7.12 and 7.13).

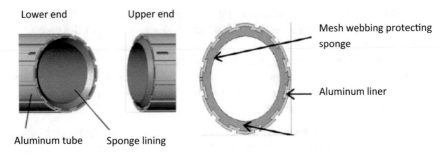

Fig. 7.11 Aluminum inner tube including the sponge (Shale et al. 2014)

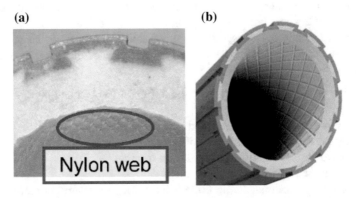

Fig. 7.12 **a** Protective nylon meshweb which protects the sponge texture and tightens the core to the sponge (Shale et al. 2014) and **b** the sponge liner (published courtesy of Baker Hughes GE)

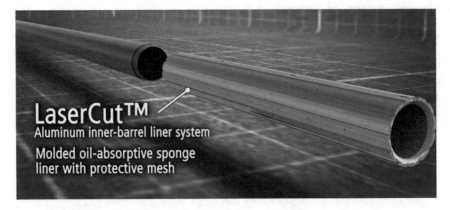

Fig. 7.13 Backside of the sponge aluminum liner including the flutes for the gas and liquid flow (published courtesy of Baker Hughes GE)

7.5.2 Challenges

Although sponge coring has been generally a rather inexpensive and direct method to improve the data obtained in oil-bearing rocks, preserve the core and mitigate the invasion (Lingen et al. 1997), it had some challenges chronologically. The chronological drawbacks are as follows (Shale et al. 2014):

- Significant friction of the core entry into the inner tube (particularly in jamming-prone formations, e.g., fractured formations), may damage/tear or roll up the sponge and/or result in rock mechanical damage and additionally loss of the saturation data.
- A fraction of oil may be expelled and lost, particularly through the annulus between the sponge and the core, by the expanding gas during tripping as the pressure drops.
- The maximum core length is usually limited to 30 ft.

Thus, the system has been reformed by the newly developed system (named SOr) to respond to the challenges as follows (inferred from Shale et al. 2014):

- Recently, the foam texture of the sponge has been greatly enhanced by developing and adding a protective mesh web. This reduces the probability of sponge damage and core jamming. The mesh web tightly makes the cut core become fit inside the sponge. Therefore, during tripping and the pressure drop, this tight contact in the new system prevents the migration of the fluids through the annulus between the sponge and the core. Then, the fluids can rise from the backside of the sponge liner up through the flutes and escape the pressure relief valve.
- The brine that is inserted into the sponge using the vacuum pump may be less than expected. Therefore, by utilizing a saturation unit, the maximum brine saturation in the sponge is ascertained.

References

Al Housani, H., F. El Wazeer, A.E. Aly, H. Al-Sahn, and S. Xu,. 2012. Using Residual Oil Saturation (Sor) to Reduce Uncertainty in Reservoir Modeling. A Case Study, SPE 161531-MS. Presented at the Abu Dhabi International Petroleum Conference and Exhibition, Abu Dhabi, UAE, November 11.

Dennis, D. (1999) Development Planning with Low Invasion Coring and Outcrop Studies, SPE 1199-0054 JPT. *Journal of Petroleum Technology*

Lingen, P.P.V., P.J.C. Schermer, C.P.J.W. Kruijsdijk, and P. Sonet. 1997. Analysis and Modeling of Flow Processes During Sponge Coring, SPE 38690 MS. Presented at the SPE Annual Technical Conference and Exhibition, San Antonio, Texas, October 5–8.

Park, A., and C.A. Devier. 1983. *Improved Oil saturation Data Using Sponge Core Barrel*, SPE 11550 MS. Presented at the SPE Production Operations Symposium. Oklahoma City, Oklahoma, 27 February–1 March.

Rathmell, J.J., R.R. Gremley, and G.A. Tibbitt. 1994. Field Applications of Low Invasion Coring, SPE 27045 MS. Presented at the SPE Latin America/Caribbean Petroleum Engineering Conference, Buenos Aires, Argentina, April 27–29.

Rathmell, J., L.K. Atkins, and J.G. Kralik. 1999. *Application of Low Invasion Coring and Outcrop Studies to Reservoir Development Planning for the Villano Field*, SPE 53718 MS, Presented at Latin American and Caribbean Petroleum Engineering Conference, Caracas, Venezuela April 21–23.

Shale, L., S. Radford, T. Uhlenberg, J. Rylance, A. Kvinnesland, and C. Rengel. 2014. *New Sponge Liner Coring System Records Step-Change Improvement in Core Acquisition and Accurate Fluid Recovery,* SPE 167705 MS. Presented at the SPE/EAGE European Unconventional Resources Conference and Exhibition, Vienna, Austria, February 25–27.

Skopec, R.A., and G. McLeod. 1997. Recent Advances in Coring Technology: New Techniques to Enhance Reservoir Evaluation and Improve Coring Economics, PETSOC 97-11-02. *Journal of Canadian Petroleum Technology.*

Whitebay, L., J.K. Ringen, L.V. Puymbroek, L.M. Hall, and R.J. Evans. 1997. *Increasing Core Quality and Coring Performance Through the Use of Gel Coring and Telescoping Inner Barrels*, SPE 38687 MS. Presented at the SPE Annual Technical Conference and Exhibition, San Antonio, Texas, October 5–8.

Chapter 8
Mechanical Core Damage Investigation and Mitigation

8.1 Introduction

To ensure reliable core analysis results, in addition to the invasion-related core damage (discussed in Chap. 7), the least possible mechanical damage should occur to the core. Mechanical core damage is defined as a permanent change in rock properties which is not reversible by restoring the rock back to in situ conditions. This damage can occur during core-drilling and while tripping of the core barrel to the surface (Santarelli and Dusseault 1991; Bouteca et al. 1994; Hettema et al. 2002; Rosen et al. 2007). During the drilling, the core undergoes severe physical stresses which can cause mechanical damage. Next, during the tripping to the surface, the core sample undergoes severe pressure and temperature changes which induce tensile stresses.

In order to investigate the source of damage, as will be discussed in this chapter, it is first greatly prominent to conduct rock mechanical simulation studies to investigate the extent and possible occurrence of the damage. Next, some measures are required by optimizing the drilling and tripping parameters to minimize the damage.

8.2 During Core-Drilling

The mechanical damage to the core can occur during core-drilling/drill-out operation fall into induced fractures and excessive vibrations.

© Springer International Publishing AG, part of Springer Nature 2018   123
R. Ashena and G. Thonhauser, *Coring Methods and Systems*,
https://doi.org/10.1007/978-3-319-77733-7_8

8.2.1   Induced Fractures

The operational parameters (already discussed in Sect. 9.2) should be designed and adjusted in a way that coring is not performed in an already induced-fractured rock (induced, e.g., due to too much WOB). Induced tensile fractures cannot only cause jamming but also adversely alter core analysis results. Thus, to prevent this, it is recommended to simulate the induced stresses ahead the core bit using geomechanical modeling (Santarelli and Dusseault 1991; Bouteca et al. 1994; Hettema et al. 2002). The modeling can be analytical-numerical such as axisymmetric 3D models of the bit–rock interaction. As an example, Fig. 8.1 shows the simulation of stress distribution contours for a parabolic core bit.

Among the operational parameters, it is of great importance during core-drilling to know the maximum permissible WOB that can induce fractures ahead the bit. Therefore, it can be notified to the driller that it is not exceeded. Otherwise, the formation ahead the bit would experience fractures prior to being cored. This may lead to probable jamming during the core entry into the inner tube (which can be identified by changes in the operational drilling parameters as discussed in Sect. 9.2). Considering the Mohr Coulomb's failure envelope, the maximum WOB during coring operations (above which the tensile failure of the core occurs) is found by (Santarelli and Dusseault 1991; Hettema et al. 2002):

$$\text{Max WOB} < \frac{1}{2} A_{\text{core}} (UCS + \tan(\varnothing) \times 0.5P), \qquad (8.1)$$

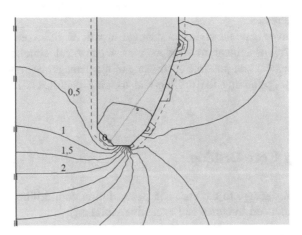

Fig. 8.1 Contours of mean principal stress induced by the axisymmetric model of a parabolic core bit (with WOB of 10 tonne, vertical stress of 7400 psi, minimum and maximum horizontal stresses of 6380 psi, MW of 13.34 ppg, at the True Vertical Depth (TVD) of 2540 m). The rock material has been assumed linear elastic. The curves with the values of 0.5–2 show the stress ratios (Hettema et al. 2002)

where UCS is the cohesion, $\emptyset$ is the frictional angle, ΔP is the differential or overbalance pressure [psi], and A_{core} is the area of the core [in^2]. The factor of 0.5 has been used by Hettema et al. (2002) to consider the effect of mud invasion on the reduction of the real overbalance pressure to the assumed 50%. It is noted that the invasion of the mud into the formation reduces the mud overbalance pressure and the confining stress support to the core.

In addition to the operational parameters, the next significant origin of induced fractures during coring occurs due to core-jamming or stuck in the inner tube which will be discussed in the next chapter.

8.2.2 Vibration

Particularly in deviated boreholes, the vibration of the bit and the BHA can potentially cause mechanical core damage (including induced fractures which can cause jamming). The BHA and the core may undergo a combination of axial, lateral, or torsional vibrations. Therefore, prior to coring, considering the formation properties, the optimization of the design of the coring BHA, the bit, and also the operational parameters are recommended to prevent excessive vibrations. Thus, usually, the coring BHA are made greatly stiff using stabilizers, which should be 1/32-in. under-gauged and positioned at 30 ft spacing (Khan et al. 2014). For long core barrels, the number of stabilizers should not exceed three (one near-bit sta-bilizer, one in the middle and one on the top) and a short sub (~ 1 m) replaces the stabilizer for each outer tube joint without stabilizers. Greater number of stabilizers can potentially lead to core barrel/outer tube stuck and termination of the coring operations. The core bit should be anti-whirl and its properties should be optimized considering the formation properties. The operational parameters, particularly the WOB and the rotary speed, (discussed in Sect. 9.2) should be optimized using geomechanical simulations to prevent critical vibrations (inferred from Hettema et al. 2002).

8.3 During Tripping

To prevent mechanical damage during tripping, it is crucial to (1) optimize the tripping rate to prevent excessive depressurization of the core during the retrieval and (2) to use an optimal core barrel length particularly in case of highly deviated or horizontal boreholes and (3) minding some operational measures.

8.3.1 Tripping Rate

When the core is being retrieved from the bottom-hole to the surface, due to the pressure and temperature drop, the pore fluids expand and are expulsed out of the core. Fast decompression of cores may not allow sufficient diffusion time for the pore fluids in the center of the core to dissipate than in the annulus. This is the most severe in cores with ultra-low-permeability or with very low-permeability mud cake (McPhee et al. 2015). This induces significant pore pressure difference within the sample (inferred from Schmitt and Li 1994; Holt et al. 2000). As this may cause tensile failure and thus the creation of microfractures within the sample (inferred from Schmitt and Li 1994; Holt et al. 2000), it causes adverse alteration of the core properties, particularly the rock and mechanical ones (e.g., porosity, permeability, compressive and tensile strength, Young's modulus). This phenomenon is attributed to the gas expansion and viscous forces created by that (Norrie et al. 2002; Rosen et al. 2007). A very specific characteristic of these fractures is the fact that they initiate within the central part of the core rather than near its boundary (Bouteca et al. 1994; Ashena et al. 2018a). Thus, no apparent mechanical damage may be visible in the retrieved cores whereas they may have undergone severe mechanical damage. In addition, the microfractures may propagate, connect, and convert into macro fractures, which may even cause core-jamming during tripping. An example of CT scans from a damaged tight core due to the decompression is shown in Fig. 8.2.

In order to prevent the tensile failure and the creation of microfractures in the cores during tripping, first, the maximum allowable tripping rate/speed must be determined. Then, the tripping rate should be adjusted not to exceed the maximum

Fig. 8.2 CT scan images showing microfractures in a core (Mcphee et al. 2015; Zubizarreta et al. 2013; Byrne et al. 2015)

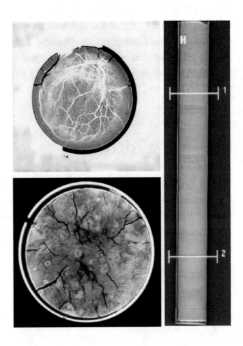

Table 8.1 A generic core trip schedule for conventional coring in shale-gas reservoirs (Zubizarreta et al. 2013)

Depth (TVD, m)	Trip rate
Bottom—Top of BHA at surface	0.45 m/s (1 min/stand)
Top of BHA at surface—150	0.075 m/s (6 min/stand)
150—Surface	0.05 m/s (9 min/stand)

A drill-pipe *stand* in conventional coring consists of three joints with total length of about 27 m

Table 8.2 Another generic core trip schedule for conventional coring in shale-gas reservoirs

Depth (TVD, m)	Trip rate
Bottom—400	0.45 m/s (1 min/stand)
400–100	0.075 m/s (6 min/stand)
100–surface	0.045 m/s (10 min/stand)

allowable value at each depth. The industry has had some attempts in this way. To prevent exceeding the maximum allowable rates, several generic or rule of thumb methods are available in the industry or standard organizations. For example, in API RP 40, (1998), it is stated: '*The core barrel should be brought to the surface smoothly. During the last 500 ft (≈150 m) the core should be surfaced slowly to minimize gas expansion that can severely damage unconsolidated cores if the pressure is reduced too quickly*'. Several companies have developed their own core tripping schedules which are based on rules of thumb or experience (McPhee et al. 2015). Two generic schedules for the tripping rates are shown in Tables 8.1 and 8.2. These and similar other schedules for core tripping are too general and may not either protect the mechanical integrity of the core during tripping or conversely, may cause greater than enough tripping times. Therefore, they are not appropriate enough for an engineering application.

To develop an engineering method for modeling the core decompression rate during retrieval, recently, some works have been conducted using poroelastic modeling (Santarelli and Dusseault 1991; Hettema et al. 2002), fluid flow by Computational Fluid Dynamics software (Zubizarreta et al. 2013; Byrne et al. 2015) and Finite Element modeling (Hoeink et al. 2015). In all the previous works, either the maximum allowable tripping rates have not been evaluated, or if evaluated by in Hettema et al. (2002), not all the necessary factors have been considered and modeled, e.g., the mud cake pressure drop was just assumed constant and thus not realistically modeled, the swabbing effects were neglected. Therefore, to obtain a comprehensive method, a Thermo-Poro-Elastic (T-P-E) approach have been used by Ashena et al. (2018b) to evaluate the maximum allowable tripping rates by considering all the effects (including the neglected ones in the literature). Finally, the maximum allowable rates have been evaluated. In this approach, the following assumptions have been made.

(1) The derivation is made within the linear T-P-E framework.
(2) Core drainage or pore fluid diffusion occurs only along the core radius. Therefore, axial flow with respect to the radial flow is neglected.
(3) Core pulling occurs at a constant rate for each depth element.
(4) The stress state is hydrostatic mud pressure, i.e., the confining pressure applied to the core is the mud pressure.
(5) The sample pore pressure is initially assumed equal to the bottomhole mud pressure.

Having developed a T-P-E geomechanical model and using the above assumptions, the following equation has been found by Ashena et al. (2018b) which can directly find the maximum allowable tripping rate, in m/s ($V_{c,trip}$):

$$V_{c,trip} = \frac{T.S. - K_{swab}}{4.7gMW\frac{R_c^2}{\eta}(1 + C_{mc}) + C_{thermal} + C_{swab}},\qquad (8.2)$$

where

K_{swab} and C_{swab}	the swab coefficients (must be calculated).
MW	mud weight (Kg/m³).
R_c	core radius (m).
C_{mc}	mud cake coefficient (must be calculated).
$C_{thermal}$	thermal coefficient (must be calculated).

Using Eq. 8.2 and the input data for a typical core as in the Appendix (but with using different permeability values and fluid type in the core), Ashena et al. (2018b) have found the tripping rates, as shown in Fig. 8.3.

Using the modeling results, it is generally inferred that: (1) the maximum allowable tripping rates decrease with decreasing permeability or hydraulic diffusivity[1] (with the same fluid content), and (2) the maximum allowable tripping rate be reduced for the last few hundreds near the surface (except when the core fluid is water). Ashena et al. (2018b) show that water-bearing cores can be safely tripped as quickly as wireline (~ 1.524 m/s), even with permeabilities as low as 0.01 mD. For gas and oil-bearing cores, it is different from water-bearing ones as the fluids expand with pressure drop. For gas cores with the permeability of 0.01 mD (or the hydraulic diffusivity at surface conditions of $10^{-6}\frac{m^2}{s}$), tripping can be safely done via the conventional pipe tripping. Very tight shale-gas cores (permeability in the order of 0.001 mD and diffusivity $\sim 10^{-8}\frac{m^2}{s}$), should be tripped extremely slowly to have safe retrieval, which it is not economic (Fig. 8.3b). Oil-bearing cores with the same rock properties as the gas-bearing one must be tripped with lower rates

---

[1]Hydraulic diffusivity (η) is found by: $\eta = 9.869 \times 10^{-13}\frac{K}{\varphi\mu C_t}$, where K is permeability (mD), φ is porosity, μ is viscosity (cp); and C_t is the total compressibility (1/Pa).

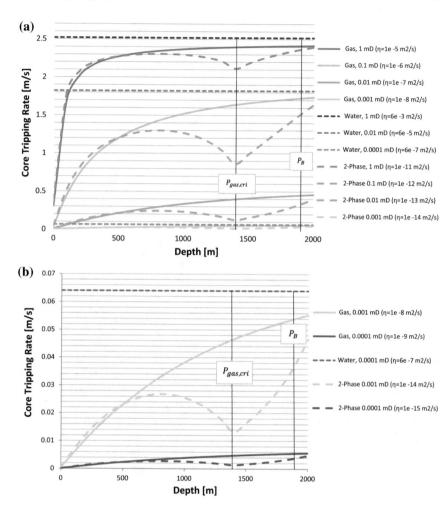

Fig. 8.3 Maximum allowable tripping rate for gas, oil, and water-bearing cores with different permeability and hydraulic diffusivity values: **a** for 0.01–1 mD, **b** for 0.0001–0.001 mD (Ashena et al. 2018b)

particularly in the vicinity of the interval between the bubble point (here 1910 m) and the gas saturation point. This is attributed to the effect of sudden gas liberation and expansion out of the core at the bubble point which pushes the oil out and applies significant viscous forces across its pore throats, which can potentially cause microfractures. This is termed by Santarelli and Dusseault (1991) as the permeability blockage. In general, with the same rock properties and conditions, it is the quickest and easiest to trip the water-bearing cores of all; then, it is easier to trip the gas-bearing cores than the oil-bearing ones.

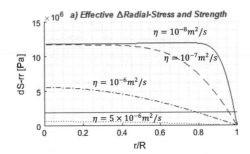

Fig. 8.4 The induced radial stress, y-axis, versus the radial location from the center, x-axis ($0 \leq$ r/R ≤ 1) using wireline tripping rate (Ashena 2017). The red line represents the tensile strength of the core

The safe tripping rate can be also found using another T-P-E approach which is based on controlling and maintaining the induced stresses less than the tensile strength of the rock. In this approach, the induced stresses are the output and the maximum tripping rate can be found in a trial and error manner (Ashena 2017). Figure 8.4 shows the induced stresses in a typical gas-bearing core initially located at 4000 m which is tripped with the wireline tripping rate. In case the induced stresses, e.g., the radial, exceed the tensile strength of the core (shown by the red line in the same figure), it signifies an unsafe retrieval. As it can be seen, if the hydraulic diffusivity of the gas core is greater than 5×10^{-6} m²/s (typical of non-tight sandstones), it can be retrieved safely to the surface with quick wireline speed. Using these simulations (Ashena 2017), gas shale cores with the hydraulic diffusivity of 10^{-8} m²/s (typical of gas shales) cannot be retrieved safely to the surface unless with extremely low uneconomic rates which are not economic. Therefore, for such cores, controlling the tripping rates cannot be used to prevent mechanical damage and microfractures and pressure coring is recommended (refer to Chap. 11).

8.3.2   Core Barrel Length

In a deviated or horizontal well, when the inner tube (containing the core) is being tripped to the surface, it undergoes bending stresses. Depending on the stresses and the rock strength, the core may be fractured. Thus, it is of great importance to optimize the core barrel length considering the wellbore geometry (inclination angle, dog leg severity, hole size, etc.) depending on the formation, to prevent excessive bending leading. Assuming the core to be linear elastic, the maximum core barrel length (L_{max}) for a rather consolidated formation is determined by (Hettema et al. 2002):

$$L_{\max} = 2\sqrt{(D_{\text{hole}} - OD_{\text{OT}})\left[\frac{3438}{\text{DLS}} - (D_{\text{hole}} - OD_{\text{OT}})\right]}, \qquad (8.3)$$

where

DLS   Dog leg severity (°/30 m).
D_{hole} Hole diameter (m).
OD_{OT} Outside diameter of the outer tube (m).

It can be inferred from the above equation that the larger the DLS value is, the less the maximum core barrel length has to be. For example, for conventional coring in 6/in. and 8½-in. hole size, using Table 15.1 we can see that, respectively, the outer tube with OD of 4¾ and 6¼-in. can be used. The optimal core barrel length has been evaluated for both cases as in Fig. 8.5. We can infer that shorter barrels should be used for slimmer hole sizes.

Another point to consider about the length is for coring in unconsolidated unstable formations. In such formations, it is not possible to obtain long cores without disturbance. To mitigate this mechanical disturbance and the core compression damage, the inner tube should be completely filled (equivalent to highest possible coring efficiency). One way to increase the possibility of complete filling is by using short core barrels, e.g., 30–60 ft (~ 10–20 m) (inferred from McPhee et al. 2015). It should be noted that particularly for long cores, to prevent core damage and jamming, the type of the drilling system greatly matters: the top-drive drilling system is preferred over the Kelly system (as is explained in the next section).

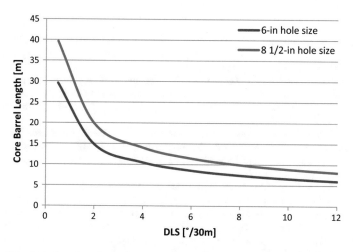

Fig. 8.5 A typical core barrel length design versus DLS variations. The blue and red lines, respectively, correspond to 8½-in. and 6-in. hole sizes with the same outer tube OD of 6¾-in.

8.3.3 Operational Measures

During coring, there are some operational measures to mind:

Using the top-drive system, coring can be continued nonstop for three joints (~ 90 ft) prior to making any connections. In addition, there is no need to raise the drill string; thus, the core is not broken. Using the Kelly/rotary-table system, for each pipe connection (every ~ 30 ft), to raise the Kelly (and thus the drill string) and set the slips, first, the core is inevitably broken. Following the connection, coring can be continued. This may cause core damage or even jamming. To prevent this; therefore, the top-drive is preferred over the Kelly one.

In case the Kelly system is only available and used, the rotary table should not be used for backing-out the drill string because it can damage the core sample inside (inferred from McPhee et al. 2015).

During the POOH in conventional coring, to prevent core damage, it is strongly recommended to set the slips gently to prevent core damage, or even jamming and loss (inferred from McPhee et al. 2015).

There are some other handling measures especially recommended for long coring. For example, the inner tubes should be equipped with Nonrotating Inner Tube Stabilizers (NRITs) to prevent excessive disconnection torques (which can damage the contained core). Next, using cradles, special care should be paid to carry the recovered inner tubes from the drilling floor to the ground (for more information, refer to Chap. 14).

References

American Petroleum Institute API. 1998. *Core Analysis*, recommended practices-40 (RP-40).

Ashena, R. 2017. *Optimization of Core Tripping Using a Thermoporoelastic Approach*. Ph.D. diss., Submitted to Montanuniversität Leoben.

Ashena, R., W. Vortisch, G. Thonhauser, V. Rasouli, and S. Azizmohammadi. 2018a. Optimization of Core Retrieval Using a Thermo-Poro-Elastic (T-P-E) Approach. Published in *Journal of Petroleum Science and Engineering*, August, 167: 577–607, https://doi.org/10.1016/j.petrol.2018.04.041.

Ashena, R., G. Thonhauser, A. Ghalambor, V. Rasouli, and R. Manasipov. 2018b. *Determination of Maximum Allowable Safe Core Retrieval Rates*, SPE-189480. Presented at the SPE International Conference and Exhibition on Formation Damage Control held in Lafayette, Louisiana, USA, February 7–9.

Bouteca, M.J., D. Bary, J.M. Piau, N. Kessler, M. Boisson, and D. Fourmaintraux. 1994. *Contribution of poroelasticity to Reservoir Engineering: Lab Experiments, application to Core Decompression and Implication in HPHT Reservoir Depletion*, SPE 28093-MS. Rock mechanics in Petroleum Engineering, Proceedings SPE/ISRM Eurock, Balkema, Rotterdam, 525.

Byrne, M., I. Zubizarreta, and Y. Sorrentino. 2015. *The Impact of Formation Damage on Core Quality*, SPE 174189-MS. Presented at the SPE European Formation Damage Conference and Exhibition, Budapest, Hungary, June, 3–5.

Hettema, M.H.H., T.H. Hanssen, and B.L. Jones. 2002. *Minimizing Coring-Induced Damage in Consolidated Rock*, SPE-78156-MS. Presented at the SPE/ISRM Rock Mechanics Conference, Irving, Texas, October 20–23.

Hoeink, T., W. Van Der Zee, and S. Arndt. 2015. *Optimal Core Retrieval Time for Minimizing Core Decompression Damage*, SPE 176943-MS. Presented at the SPE Pacific Unconventional Resources Conference and Exhibition, Brisbane, Australia, November 9–11.

Holt, R.M., M. Brignoli, and C.J. Kenter. 2000. Core Quality: Quantification of Coring-induced Rock Alteration. *International Journal of Rock Mechanics and Mining Sciences* 37 (6): 889.

Khan, A.M., J. Onezime, M. Diaa, T. Amos, M. Pointing, and V. Vyas. 2014. *Successful Coring in 8.5-in Hole Section Using Anti-Jamming Technology in Southern Iraq*, SPE 172099-MS. Presented at the Abu Dhabi International Petroleum Exhibition and Conference, Abu Dhabi, UAE, November 10–13.

McPhee, C., J. Reed, and I Zubizarreta. 2015. *Core Analysis: A Best Practice Guide*, 52–57. Published by Elsevier. ISBN-13:978-0444635334.

Norrie, A.G., S. Akram, and M.R. Niznik. 2002. *North Sea Coring Record Attained by Step Change Application Modelling*, SPE 74509. Presented at the IADC/SPE Drilling Conference, Dallas, Texas, February 26–28.

Rosen, R., B. Mickelson, J. Fry, G. Hill, B. Knabe, and M. Sharf-Aldin. 2007. *Recent Experience with Unconsolidated Core Analysis*. Presented at the International Symposium of SCA, Calgary, Canada, September 10–12.

Santarelli, F.J., and M.B. Dusseault. 1991. *Core Quality Control in Petroleum Engineering*. Balkerna, Rotterdam: Published in the Rock Mechanics as a Multidisciplinary Science.

Schmitt, D.R, and Y.Y. Li. 1994. *Determination of the Microcrack Tensor in Rock: Evaluation of Coring Induced Damage*, ARMA-1994-0443. Presented at the 1st North American Rock Mechanics Symposium, Austin, Texas, June 1–3.

Zubizarreta, I., M. Byrne, Y. Sorrentino, and E. Rojas. 2013. *Pore Pressure Evolution, Core Damage and Tripping Out Schedules: A Computational Fluid Dynamics Approach*. SPE 163527-MS, Presented at the SPE/IADC Drilling Conference, Amsterdam, The Netherlands, March 5–7.

Chapter 9
Jam Mitigation Coring

9.1   Introduction

One of the most common coring problems particularly in critical formations is core jamming (either at the catcher or in the inner tube) and also the risk of core loss or sliding out of the inner tube during pulling out of the hole. As was already listed in Table 5.1, most of the coring challenges lead to jamming. This originates from the:

- Formation (unconsolidated unstable, clay-containing, or fractured formations, interbedded formations, or heterogeneous sections where the formation changes and the boundaries are frequent)
- Bit design (due to the bit possible whirl and vibration)
- Improper operational parameters (e.g., low circulation rate or excessive WOB)
- BHA (when the core barrel is extremely long, particularly in highly deviated wellbores, when the inner tube is rough or its friction is high)
- Mud (muds which are highly viscous or contain Lost Circulation Materials, LCMs)
- Bottom-hole fill (the bottom-hole has junk due to inadequate hole cleaning)
- A combination of some of the above.

Although we can reduce the potential for jamming by proper selection of the bit design, the operational parameters, the BHA, the mud, and good hole cleaning, the effect of the formation may be still unpredictable.

In case of jamming, it is recommended to overpull and terminate the job resulting in an inevitable POOH, which signifies that the core barrel cannot successfully accomplish the job. However, the cost of tripping unexpectedly short cores is unacceptable. Therefore, to mitigate jamming and its problems and also prevent core sliding from the bottom, two systems (antijamming and full-closure coring) are introduced in this chapter. These systems contribute to enhancing the coring KPIs such as the coring efficiency, the core recovery, and quality (inferred from Whitebay 1986; Rathmell et al. 1994; Armagost and Sinor 1994; Al-Sammak et al. 2009; Mukherjee et al. 2015). Therefore, in this chapter, following a discussion of the operational parameters and jam detection, the antijamming and full-closure systems are described.

9.2   Operational Parameters and Jam Detection

Generally, following the design and adjustment of the coring operational parameters using offset well performance and simulations, it is crucial to monitor them during core-drilling (Guarisco et al. 2011; Keith et al. 2016). This enables the driller to detect any unprecedented downhole problems including core jamming. The main operational parameters are:

9.2.1   Rate of Penetration

Coring ROP is a greatly important parameter which is the result of the other coring parameters. Therefore, if other parameters are optimized, it would be satisfactorily great enough for coring and simultaneously low enough to prevent any undesirable vibrations and induced fractures (as it was introduced in Chap. 8, Sect. 8.2). Generally, to prevent induced fractures and jamming, the coring ROP is maintained low (lower than the drilling ROP assuming the same other conditions). The greatest possible ROP should not be the only target in coring; instead, the focus should be on optimizing the coring KPIs and retrieval of adequate length of the core with highest possible quality (Guarisco et al. 2011; McPhee et al. 2015; Keith et al. 2016).

Abrupt changes in ROP can signal a change of conditions in the operations. For example, similar to conventional drilling, a sudden variation of ROP may indicate change of the formation lithology. In case of jamming occurrence, the coring ROP may remain normal, but most likely would show a significant decrease.

9.2.2   Standpipe Pressure

Standpipe pressure is the next greatly prominent parameter during coring. It should be carefully monitored as it may vary significantly in case of coring problems and it provides the most definite sign of jamming. The following items should be minded about this parameter:

– When variations in the standpipe pressure are observed, the reason should be traced first in the surface mud circulation system and then in the downhole. In case no problem can be observed in the mud circulation system at the surface (such as leakages or plugging in pumps, lines, mud tanks, etc.), the variation must be attributed to the downhole (Table 6.1).

– A main downhole reason for standpipe pressure variation is core jamming. In conventional coring without jam-indicator sub, when jamming occurs, the standpipe pressure shows an initial increase (due to a restriction against the mud flow out of the bit). Shortly, as the jammed core pushes the bit off-bottom (which causes the ROP to drill-off or decrease significantly), the mud downhole can flow easily through the annulus resulting in less hydraulic friction and less downhole mud pressure. However, there is a time delay before the pressure drop can be transmitted up to the surface and a drop in standpipe pressure can appear and be detected. This time delay causes the driller to indeed mill away and disintegrate the already taken core. To resolve this issue and enable quicker jam detection in conventional barrels, there are two solutions. First, a core jam indicator is essentially used, which makes an immediate pressure increase signaling the jam indication (look at Fig. 9.1). Second, the bit design should be altered in such a way that the spacing between the bit ID and the lower show OD

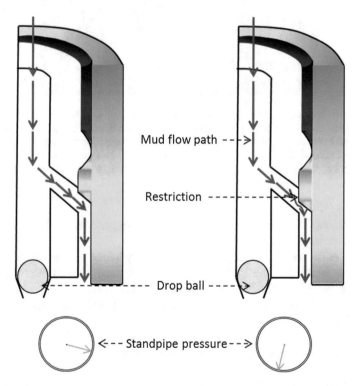

Fig. 9.1 Using core jam indicator in conventional coring to enable quick detection of jamming. In case of jamming, when the inner tube is lifted, the jam-indicator sub is moved such that it can make a restriction against the mud flow path through the annulus between the inner and outer tubes. This causes an instant increase in the standpipe pressure

(a)   **(b)**

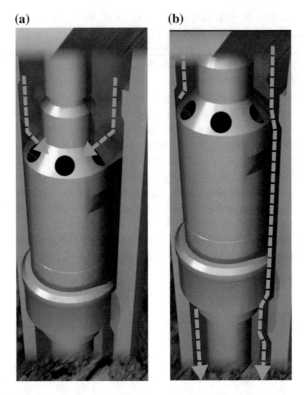

Fig. 9.2 Core jamming indication in wireline continuous coring: **a** while core drilling is in progress (prior to core jamming), **b** when jamming occurs, the inner tube is lifted up immediately (because it is hydraulic) which causes the mud flow to pass through the annulus between the inner and the outer tubes instead of through the *flow cap assembly*. This causes an instant drop of the standpipe pressure. Compare it with jamming in conventional coring equipped with a jam indicator (Fig. 9.1)

is increased. This means an increased mudflow area and thus reduced time delay for quick jam detection.

However, in wireline coring, because it is a hydraulic seating on the pressure head metal-to-metal seal, as soon as jamming occurs, the standpipe pressure drop is instantly contributing to quick detection (Fig. 9.2). This is rather an advantage for wireline coring that jam detection is possible without the necessity of using any jam-indicator sub.

Note: In conventional drilling, the driller may raise the BHA off-bottom for checking the standpipe pressure or other parameters in case of skepticism about the well conditions (e.g., possible loss of flow). However, during core-drilling, unlike conventional drilling, it is not possible because it can cause the core breakage and jamming.

9.2.3 Mud Circulation Rate

The following considerations should be made about the mud circulation flow rates during coring:

- Similar to conventional drilling, if the circulation rate during core-drilling is extremely low, the consequences are poor hole cleaning, regrinding, bit-balling, and even PDC bit burning. In turn, poor hole cleaning can also cause the core bit to become off-bottom leading to core jamming. Conversely, if the circulation rate is extremely high, it can cause wash out of the sample. Therefore, the required mud circulation rates must be optimized by conducting a hydraulics simulation (inferred from Guarisco et al. 2011). As a rule of thumb for an estimation of a minimum limit for coring circulation rate, the minimum annular velocity of 27 m/min can be considered for properly cleaning the bit face.
- About 20 m off-bottom, begin the mud circulation. At the time of dropping the ball down the hole, the circulations rate is advised to be about 150 GPM (McPhee et al. 2015). This reduces the risk of collapsing the inner tube/barrel.
- At the beginning of core-drilling, start with low circulation rates (e.g., 120–150 GPM). Depending on the hole cleaning conditions and hole size, increase the circulation rate (if necessary). Avoid extremely high rates as it may have some consequences such as lifting the bit off-bottom while coring, mud invasion, or washing-off the core (inferred from Keith et al. 2016).
- The larger the hole size is (with the same core size), the larger the magnitude of the produced cuttings and thus the required circulation rates for proper hole cleaning. For example, using conventional coring in 8½-in. hole size, the typical mud circulation rates typically range from 70 to 230 GPM. However, for the 12¼-in. hole size, for coring, the range may be from 400 to 450 GPM.
- The smaller the core size is (in the same hole size), the greater the required mud circulation rate. Therefore, for example in wireline coring (by which smaller core sizes can be obtained compared with the conventional method), comparatively larger circulation rates are required.
- Compared with drilling, the mud circulation rates in coring are less than those in drilling the same section. This is because the amount of cuttings generated by the core bit is less than that by the drilling bit.
- As mud circulation rate is not affected by core jamming occurrence, it does not provide any indication for jamming.
  Note: Circulation times should be limited in order to reduce the risk of core loss unless hole conditions dictate heavy circulation is needed. Well safety should always take precedence over core recovery (inferred from McPhee et al. 2015).

9.2.4 Weight on Bit

The following considerations should be made regarding the Weight On Bit (WOB):

– Geomechanical modeling should be conducted and the results should be considered to be aware of the WOB limits, particularly the maximum one (refer to Sect. 8.2.1, Eq. 8.1). Excessive WOB should be avoided as it causes the core bit to try to penetrate extremely quickly which adversely causes improper hole cleaning, bit-balling, and even PDC bit burning, excessive vibrations, and core jamming. Particularly in soft unconsolidated formations, strictly refrain from applying high WOB.[1] It is known that the WOB is provided by the BHA (the outer tube and drill collar in vertical wells, and the outer tube and HWDPs in directional wells). Knowing the WOB limit, we can determine the number of required drill collars (for vertical wells) or HWDPs (for directional wells).
– At the beginning of coring operations, apply low WOB (e.g., about 2–5 K-Ibs). Then, raise it until the optimal WOB (e.g., 10–20 K-Ibs) can be found. To do this, (while keeping the RPM constant) increase the WOB gradually, e.g., in steps of 1000 Ib, and monitor the ROP. Continue until the great enough ROP is attained. After reaching the required ROP (by optimizing the WOB and RPM), try to maintain it (by keeping the drilling conditions at optimal conditions) to prevent drilling-off (Guarisco et al. 2011; Keith et al. 2016). Automatic drilling can be a good method for maintaining the optimal WOB unchanged. Do not allow drilling-off during coring; otherwise, it can cause a risk of disturbing the core (inferred from McPhee et al. 2015).
   To have an idea about the typical WOB values for conventional coring, refer to Tables 15.1 and 15.2 and for wireline coring, refer to Table 15.3.
– When jamming occurs, WOB may show an increase because jamming causes the weight to apply to the whole formation rock (including the core part) instead of the washer part of the rock to be drilled. Field experience shows that in some jamming occurrences, WOB remained almost unchanged.

9.2.5 Rotary Speed

The following considerations should be made during coring about the rotary speed (RPM):

– Apply low RPM at the beginning of the operation (e.g., 30 RPM). During the operations, optimal RPM values can be found by monitoring ROP while

---

[1]In such low strength cores, in addition to the WOB control, a one-joint (not longer) core barrel is used.

increasing the RPM with constant WOB (Guarisco et al. 2011; Keith et al. 2016).

– Generally, the rotary speed in coring can be quite high. However, in fractured or unconsolidated formations, apply low rotary speeds (e.g., 30–40 RPM) to minimize the core barrel disturbance and core jamming.
– As the rotary speed is not affected or changed by core jamming occurrence, it does not provide any indication in case of jamming.

Typical values of RPM can range from 80 to 230 for conventional coring (look at the operational parameters in Tables 15.1 and 15.2). For wireline coring, due to less stiffness of the system, lower RPM values can be applied (look at Table 15.3). It is noted that the RPM values required for PDC core bits are generally lower than for PDC drilling bits, assuming the parameters to be the same.

9.2.6 Torque

Torque is one of the important parameters which is continuously monitored during the coring operations (as is the case with conventional drilling operations). Particularly, in directional or horizontal trajectories, high torque should be prevented by formulating the mud properties and BHA; otherwise, core jamming and inefficient coring may be the consequences.

When jamming occurs, the bit is lifted off-bottom resulting in less contact with the wellbore. Therefore, it causes a reduction in torque as a sign of occurrence of core jamming.

9.2.7 Observations

A combination of observations in the real-time operational parameters is essential to recognize the reasons and the possible incidents. This is shown in Table 9.1.

9.3 Antijamming Coring

To overcome unprecedented termination of coring due to jamming, a good solution is to enable coring to continue after jamming occurs. This is possible using the antijamming system, which employs a telescoping inner barrel (Fig. 9.3) to accommodate three or four jams. The telescoping inner barrel consists of a standard aluminum inner barrel/tube with maximum three thin-walled aluminum liners/sleeves fitted inside the standard tube (and a jam indicator at the top of the tool for jam detection). The standard aluminum tube is screwed into a modified upper shoe.

Table 9.1 Reasoning for observations in the drilling parameters while coring

First Sign	Second Sign	Third Sign	Reason		What to do
Standpipe pressure	Torque	ROP			
Increase	Increase	Decrease	Improper hole cleaning as the result of	Plugging of the annulus between the inner and outer tubes by LCM, junk, etc.	Increase GPM, or (if the signs still stay) cut the core and POOH
				Bit waterways plugging or cut-off	Cut the core and POOH
				Creation of circular grooves in the bit face	
				Removal of the inner tube from its place and fall down into bit	
	Decrease	Decrease	Core jamming (if a jam-indicator sub is used)		
Decrease	Increase	Decrease	Wash-out/creation of a hole in the drill pipes		
Decrease following an initial increase	Decrease	Decrease	Core jamming		

Next, the three internal aluminum liners are anchored in place to the upper shoe by aluminum shear pins. The strength of the shear pins can be varied (shear pins for the first internal liner should be configured to shear at a force lower than that of the second internal liner and so forth) and is set considerably below the strength of the formation to be cored. The formation rock must be strong enough to transmit a loading to the shear pins, otherwise, core crushing will occur without activating the tool. The system is not recommended for coring greatly weak or unconsolidated formations.

Coring proceeds as normal until a jam occurs in the first liner. As the jam develops, the core begins to pack-off the ID of the innermost liner, increasing the loading on the inner barrel. This increase in loading will continue until it overcomes the shear strength of the shear pins. When this occurs, the pins will break and allow the inner tube to slide up into the inner barrel (Fig. 9.4a). Jamming a second time releases the second line, again allowing coring to proceed (Fig. 9.4b). This process can be repeated for a third or fourth jam in the inner sleeves (Fig. 9.4c) depending upon the size of the core barrel and number of sleeves used. The coring assembly must be pulled out of the hole after the final jam (the third or fourth jam) occurs or when the barrel is full.

As antijamming system is capable of providing operators maximum four chances to core during each run, it can effectively mitigate the effects of jams by eliminating

Fig. 9.3 A schematic of antijamming telescoping inner barrel (published courtesy of Baker Hughes GE)

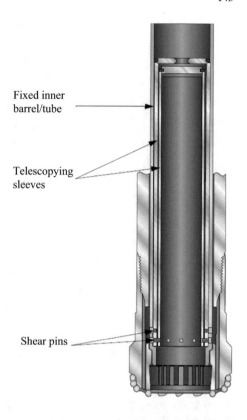

Fixed inner barrel/tube

Telescopying sleeves

Shear pins

unscheduled POOH and related non-productive time. The system therefore significantly increases the amount of core cut per run (coring efficiency) in jam-prone formations. For example, the Baker Hughes JamBuster antijamming system has increased coring efficiencies and recoveries from between 30 and 50%. Whitebay et al. (1997) compares some field cases of using antijamming and gel coring and concluded that the antijamming system showed the greatest coring efficiency and recovery in the considered interbedded jam-prone formation. Zahid et al. (2011) find out an increase in coring efficiency and core recovery in another application of the system in Pakistan. Using the offset comparison with the conventional system, they discovered that additional 1.5–2 times the length of core was recovered with antijamming system. In another investigation by Khan et al. (2014) in South of Iraq, the system showed its effectiveness to successfully core interbedded sand, shale, and limestone rocks in 8½-in. hole, with 98% coring efficiency and 100% recovery. In addition to Whitebay et al. (1997), other researchers such as Zahid et al. (2011) and Khan et al. (2014) confirmed that the antijamming system is greatly efficient in interbedded formations among jam-prone formations.

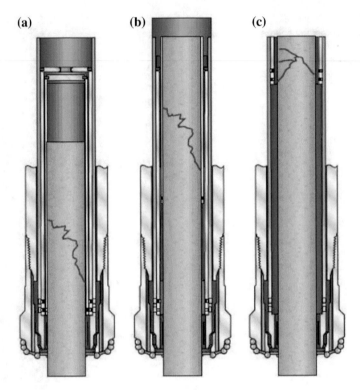

Fig. 9.4 The succession of jams in antijamming coring system (with maximum three jams): **a** core enters the first sleeve until the first jam occurs and shears the pin, **b** freed inner sleeve telescopes up the core barrel and second sleeve receives new core, **c** second jam occurs which shears the pins and releases the second telescoping sleeve. Coring continues until the third jam occurs or the inner tube is full (Zahid et al. 2011; Gehad et al. 2014)

The antijamming system has some limitations such as

– The core size may be small which may limit the advanced studies such as for SCAL studies (Zahid et al. 2011). This limitation has been recently resolved using the new systems.
– It works only if the jam occurs in an inner tube. Jams occurring at the bit or in the core catcher will not transmit loading to the shear pins, necessitating tripping the barrel.
– The system cannot be used with wireline continuous method. Because the wireline system works hydraulically, the inner tube would be detached when jamming occurs and thus coring cannot continue again (which was mentioned as one of the disadvantages of wireline continuous coring in Sect. 6.3.2).

9.4   Full-Closure Coring

The full-closure system is the other jam mitigation system which uses slight differences from the conventional or wireline coring tool to prevent core jamming, and enhance coring KPIs (including coring efficiency, core recovery and quality), particularly from critical formations such as fractured or soft unconsolidated ones (inferred from Whitebay 1986; Rathmell et al. 1994; Armagost and Sinor 1994; Al-Sammak et al. 2009; Mukherjee et al. 2015).

This system incorporates two main features: (1) a slick inner tube which reduces the friction between the core and the inner tube and thus provides slick entry of the core into the inner tube and prevents core jamming or stuck particularly at the bit throat, (2) full-closure core catcher/retainer which closes from the bottom to prevent any core loss from the bottom. The schematics of the two features have been shown in Fig. 9.5. Part-a in the figure shows the slick wall of the inner tube; part-b shows a hidden full-closure core catcher prior to activation; and part-c shows the catcher after activation with the shells/halves in closed state.

The full-closure working mechanism is explained as follows: Initially, during coring and prior to the activation of the full-closure catcher, the upper and the lower parts of the full-closure system are as, respectively, shown in Fig. 9.6, part-1 in Fig. 9.7, part-1. In this state, the dual catcher of the system is hidden behind the *smooth sleeve* shown in Fig. 9.7, part-1. At the end of coring when the core barrel is full or if core jamming occurs, a second ball (out of plastic) is dropped (from the surface or hydraulically) to be seated in its locking place (look at Fig. 9.6, part-2). The resultant mud pressure due to the sitting of the second plastic ball makes the inner tube begin lifting. This lift pulls up the smooth sleeve and uncovers the dual catcher assembly by which a heavy spring would be activated to forcefully close the two full-closure shells (Fig. 9.7, part-2). Finally, as shown in Fig. 9.6, part-3, after

(a)   **(b)**   **(c)**

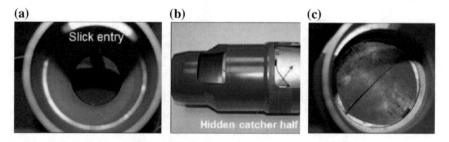

Fig. 9.5   a The slick wall of the inner tube, **b** a full-closure dual catcher prior to activation, and **c** the full-closure dual catcher after activation with the shells/halves in the closed state (published courtesy of Baker Hughes GE)

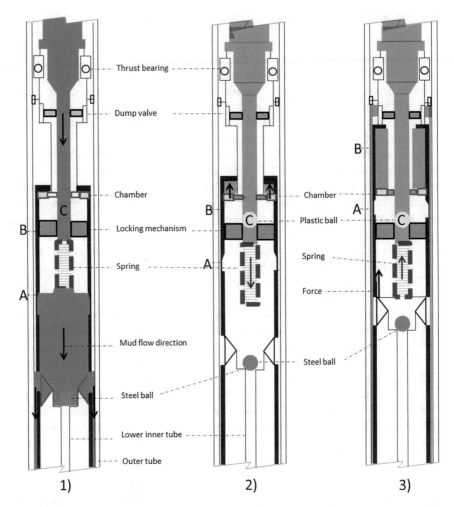

Fig. 9.6 Full-closure system (upper part): (1) before activation, (2) after activation, and (3) when complete closing and sealing has been achieved from the bottom (refer to Fig. 9.7)

the inner tube has been completely lifted (for about 8 inches), the dual catcher has fully closed or sealed the bottom of the inner tube and the *dump valve* (shown in Fig. 9.6) is activated signaling the complete sealing by the full-closure system. It is noted that points A and B in Fig. 9.6 are used to indicate the movement of the inner tube and point C represents the ball.

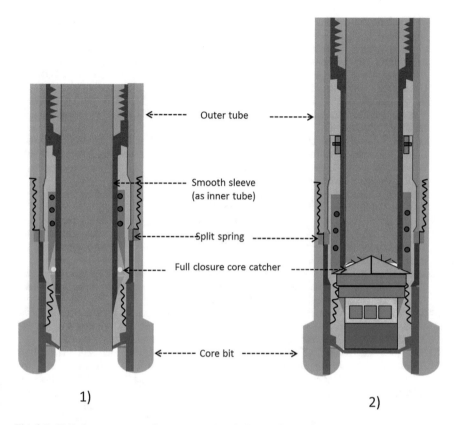

Fig. 9.7 Full-closure system (lower part): (1) during coring before activation, and (2) after activation and just prior to pulling out of the hole

The full-closure system cannot be combined with wireline continuous coring as we cannot drop-ball through the rope socket and the pressure head (this was mentioned as one of the disadvantages of wireline continuous coring in Sect. 6.3.2).

References

Al-Sammak, I., K. Ahmed, F. Al-Bous, and F. Abbas. 2009. *Coring Unconsolidated Formation-Lower Fars: A Case Study*, SPE 119918-MS. Presented at the SPE Middle East Oil and Gas Show, Bahrain, March 15–18.

Armagost, W.K., and L.A. Sinor. 1994. *The Successful Use of Anti-Whirl Technology in Conventional Coring*, SPE 27473-MS. Presented at the IADC/SPE Conference, Dallas, Texas, February 15–18.

Gehad, M.H., A.M. Salem, S.A. Shedid, M.N. Aftab, M. Ali, M.A. Reyami, W. Hazem, and F. Mohamed. 2014. *Innovative and Cost-Effective Coring Technique Extended Coring for Long Intervals of Multiple Zones with World Record-Case Histories from the UAE*, SPE

171852-MS. Presented at the Abu Dhabi International Petroleum Exhibition and Conference, Abu Dhabi, UAE, November 10–13.

Guarisco, P. Meyer, J. Mathur, R. Thomson, I. Robichaux, J. Young, C. and Luna, E. 2011. *Maximizing Core Recovery in Lower Tertiary Through Drilling Optimization Service and Intelligent Core Bit Design*, SPE/IADC 140070, Presented at the SPE/IADC Drilling Conference, Amsterdam, the Netherlands, March 1–3.

Keith, C.I. Safari, A. Aik, K.L. Thanasekaran, M. and Farouk, M. 2016. *Coring Parameter optimization-The Secret to Long Cores*, Presented at the OTC Conference, Kuala Lumpur, Malaysia, March 22–25.

Khan, A.M., J. Onezime, M. Diaa, T. Amos, M. Pointing, and V. Vyas. 2014. *Successful Coring in 8.5-in Hole Section Using Anti-Jamming Technology in Southern Iraq*, SPE 172099-MS. Presented at the Abu Dhabi International Petroleum Exhibition and Conference, Abu Dhabi, UAE, November 10–13.

McPhee, C., J. Reed, and I. Zubizarreta. 2015. *Core Analysis: A Best Practice Guide*, 52–57. Published by Elsevier. ISBN-13:978-0444635334.

Mukherjee, P., J. Peres, B.S. Al-Matar, P. Kumar, P.K. Choudhary, W.K. Al-Khamees, and M. Stockwell. 2015. *Piloting Wireline Coring Technology in Challenging Unconsolidated Lower Fars Heavy Oil Reservoir, Kuwait*, SPE 175269-MS. Presented at the SPE Kuwait Oil and Gas Show and Conference, Mishref, Kuwait, October 11–14.

Rathmell, J.J., R.R. Gremley, and G.A. Tibbitts. 1994. *Field Applications of Low Invasion Coring*, SPE 27045-MS. Presented at the SPE Latin American/Caribbean Petroleum Engineering Conference, Buenos Aires, Argentina, April 27–29.

Whitebay, L.E. 1986. *Improved Coring and Core-Handling Procedures for the Unconsolidated Sands of the Green Canyon Area, Gulf of Mexico*, SPE 15385 MS. Presented at the SPE Annual Technical Conference and Exhibition, New Orleans, Louisiana, October 5–8.

Whitebay, L., J.K. Ringen, T. Lund, L. van Puymbroeck, L.M. Hall, and R.J. Evans. 1997. *Increasing Core Quality and Coring Performance Through the Use of Gel Coring and Telescoping Inner Barrels*, SPE 38687. Presented at SPE Annual Technical Conference and Exhibition, San Antonio, Texas, October 5–8.

Zahid, S., A. Khan, and A. Khalil. 2011. *Applications of State of the Art Anti-Jam Coring System-A Case Study*, SPE 156208. Presented at the SPE/PAPG Annual Technical Conference, Islamabad, Pakistan, November 22–23.

Chapter 10
Oriented Coring

10.1 Introduction

Following coring and retrieval of the core sample, knowing its exact orientation with respect to the formation bedding contributes to better understanding the geological structure of the reservoir. Knowing the structure, in turn, helps to identify the most efficient recovery methods. However, this knowledge is not conventionally available unless oriented coring is utilized. Therefore, application of this method is mainly recommended in complex reservoir structures in the drilling and production phases (inferred from Brindley 1988; Laubach and Doherty 1999; Dennis et al. 1987).

Therefore, in this chapter, first, the oriented coring is described. Next, the components and the arrangement of the required tools are depicted, followed by the explanation of different types of the survey tools.

10.2 Description

In general, oriented coring utilizes a combination of mechanical, magnetic, and gravimetric measurements to identify the core orientation with respect to the formation dip and strike. Using the triaxial magnetic and gravimetric survey tools, the inclination angle and the azimuth are continuously measured along the core length (Nelson et al. 1987; Brindley 1988). Simultaneously, the core is mechanically marked along its length during its cutting to make a reference line (for the survey equipment). This is accomplished using three scribe knives: the main knife makes the main scribe line/groove, (called 'Main Reference Line (MRL)')[1] with the other two for the subsidiary scribe lines. Using this combination of measurements and the

---

[1]Also called *Main Reference Line (MRL)* or *Master Orientation Line (MOL)*.

© Springer International Publishing AG, part of Springer Nature 2018

R. Ashena and G. Thonhauser, *Coring Methods and Systems*,

https://doi.org/10.1007/978-3-319-77733-7_10

geological study of the sample at the surface, generally the core original orientation with respect to the dip and strike of the formation bedding and faults and also microfractures can be known, along the depth (inferred from Nelson et al. 1987; Brindley 1988; Laubach and Doherty 1999; Rourke and Torne 2011). This means that finally the three-dimensional position of the core in the formation can be characterized, which enables the three-dimensional structural modeling of the formations.

Specifically, the following parameters can be measured using this system:

1. Hole azimuth,
2. Hole inclination angle or deviation,
3. Formation dip angle,
4. Formation strike angle,
5. Microfracture azimuth (or direction),
6. Azimuth (or direction) of stresses (also called *stress orientation*),
7. Formation anisotropy.

   - Direction of permeability[2]
   - Direction of fluid migration
   - Direction of the formation deposition.

A schematic of the oriented coring and its downhole measurements through the underground bedding planes has been shown in Fig. 10.1. It is noted that oriented coring can be used both with conventional and wireline continuous methods, and it can be potentially combined with other systems such as gel or sponge coring. Currently, a combination of many coring systems is not popular for clients.

10.3   Tool Arrangement and Components

It can be inferred from the previous discussions that modifying the core barrel assembly for oriented coring is not difficult. The schematics of the oriented tool arrangement including its components for conventional and wireline continuous methods have been shown in Figs. 10.2 and 10.3, respectively. To achieve this, the following accessories must be added (according to the corresponding patents in 15.3):

The Survey Tool
The survey tool/equipment is an essential component of the oriented coring system, placed at the upper part of the coring assembly. The arrangement of the components of the survey tool is shown in Fig. 10.4. The tool consists of the spear point or rope socket for retrieval by the wireline and overshot; the centralizers to maintain the tool centered while taking the survey: the bull plug which is the swivel bearing; the

---

[2]Especially, in formations consisting of fluvial deposits.

Fig. 10.1 Schematic of
oriented coring (published
courtesy of Baker Hughes
GE)

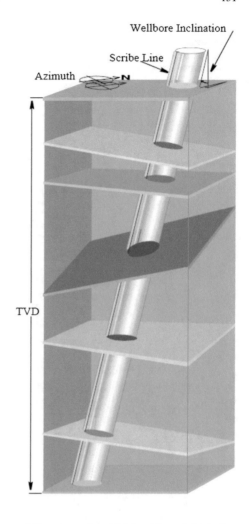

pressure barrel which houses the probe; and the spacer bar which adds weight to the
tool. The survey is taken by the probe. It is equipped with (1) the magnetometers (in
order to measure the north direction and thus, the azimuth angle with respect to it)
and (2) the gravitational accelerometers in order to measure the inclination angle of
the core while it is being cut. This is done by continuously measuring the offset
angle from the Main Reference Line (MOL). For a reliable measurement, it is
required that the survey tool mark (or the tool face) be mechanically aligned with
the main reference line (MOL) of the scribe knife. It is noted that the survey tool
should be latched properly within the non-magnetic tubes to protect it from any
undesirable mechanical or hydraulic interference. It should be also centralized using
non-magnetic centralizers.

When the oriented tool is combined with the conventional system, the survey
probe (e.g., gyro) is placed over the inner tube and is only surrounded by the

Fig. 10.2 The arrangement
of the oriented core barrel
(modified from Fig. 1 of Hay
et al. 1990)

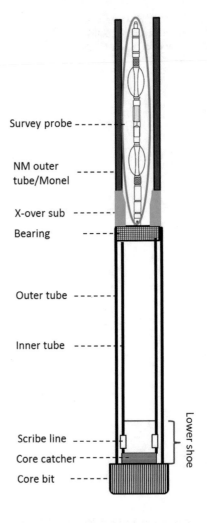

Survey probe

NM outer
tube/Monel

X-over sub

Bearing

Outer tube

Inner tube

Scribe line

Core catcher

Core bit

Lower shoe

non-magnetic outer tubes (as shown in Fig. 10.2), it is also run in the hole or retrieved as part of the coring assembly via conventional pipe trips. When the oriented tool is combined with the wireline continuous coring,[3] the probe is located in the upper part of the core barrel and is surrounded by both the non-magnetic inner tubes and outer tubes (as shown in Fig. 10.3). Retrieval of the survey tool is done via the wireline assembly by attaching to the spear point, on each occasion the inner tube is retrieved to surface.

---

[3]Currently, many coring companies cannot run the oriented coring for the wireline method as it is not feasible with their current kits. However, it would be potentially possible with some engineering input or modification of the wireline continuous system to accommodate the oriented mode system. Such a system is expected to face the limitation of use in high inclination angles (e.g., for angles > 25°).

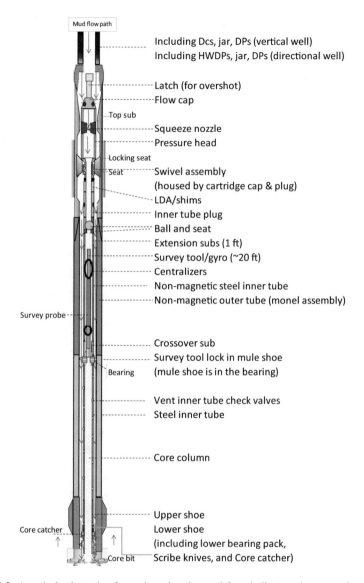

Fig. 10.3 A typical schematic of an oriented coring tool for wireline continuous coring

Alignment Tools

To ensure that the survey tool reference line is aligned with the main reference line (MOL) of the scribe knife, some alignment tools are required as shown in Fig. 10.5. Some of these tools have been shown in Fig. 10.6.

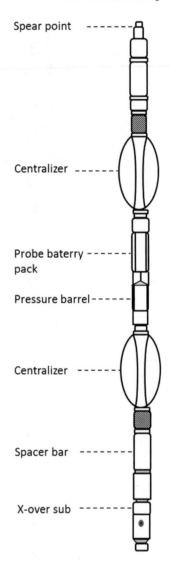

Fig. 10.4 The arrangement of a typical survey tool (inferred and modified from Figs. 2 and 3 of Brindley and Sperry-Sun 1988; Fig. 1 of Hay et al. 1990)

Spear point

Centralizer

Probe baterry pack

Pressure barrel

Centralizer

Spacer bar

X-over sub

Non-magnetic (NM) Tubes
Non-magnetic outer tubes are required in the upper part of the core barrel in order to prevent magnetic interference with the readings of the survey tool. In case of wireline oriented coring, the upper inner tubes must be non-magnetic.

Non-magnetic Extension Sub
It may be necessary to add a non-magnetic extension outer tube/sub above the inner tube to extend it, e.g., for one foot, to the top of the core barrel.

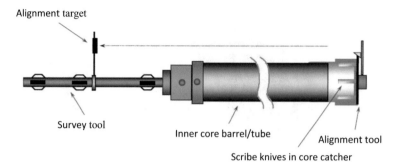

Fig. 10.5 The location of the survey in oriented coring with conventional core barrel (modified from Fig. 4 of Skopec et al. 1992)

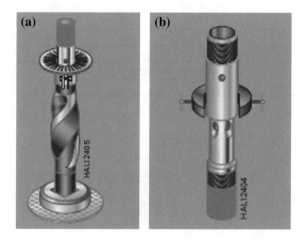

Fig. 10.6 a The *swivel assembly clamp*, which is used during the alignment of the primary scribe knife to the survey tool, **b** *the inner tube protractor*, which is also used during alignment of the primary scribe knife to the survey tool. It enables the measurement of the angle between the Main Scribe Line of the knife and the survey tool reference line (published courtesy of Halliburton)

Scribe/Marking Knives

Three inwardly-facing scribe knives (also called *marking blades/tool*) are added to the bottom of the inner tube (i.e., in the lower shoe) to mark three scribe lines/ grooves on the core as it enters the tube. Out of these three, the main knife marks the reference/main scribe line, the rest two are asymmetrically positioned (130–150° from the reference knife) to contribute to the certainty of the determination. This also helps to make the main reference scribe (line) be identifiable either from the top and bottom of the core after its retrieval to the surface. It is critical before coring to ensure that the scribe knives are adjusted such that the reference scribe

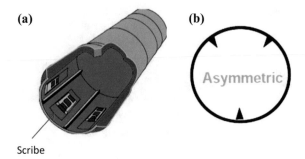

Scribe

Fig. 10.7 a Typically oriented scribe knives, **b** asymmetric arrangement of knives with 72°, 144°, and 144° angle difference between knives (published courtesy of Baker Hughes GE)

Table 10.1 Type and size of scribe knives used to mark scribe lines on the core based on formation type

Formation type	Knife type	Line size (mm)
Hard	Small	1
Medium	Medium (or standard)	3
Soft	Large	5

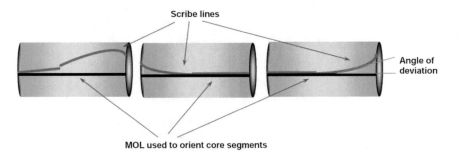

Fig. 10.8 Main reference line (MRL) or master orientation line (MOL) indicated by the survey tool, used to identify the angle of deviation and azimuth

(line) is aligned with the survey tool. This enables the alignment of the reference line by the survey tool and the main reference line by the main knife.

A schematic of the scribe knives with their asymmetric arrangement is shown in Fig. 10.7. They can be made of tungsten carbide or diamond. Depending on the formation strength and the size of the line/groove marked by the designed scribe, knives differs (Table 10.1). Typical trends of the scribe lines on the core samples are shown in Figs. 10.8 and 10.9.

Fig. 10.9 Main reference
line or master orientation line
(MOL). **a** Proper quality
reference line/groove for a
vertical well, **b** poor reference
line, and **c** a typical reference
line in a directional well

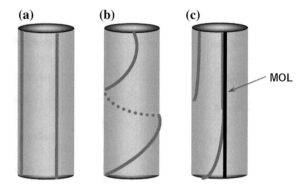

10.4   Types of Survey Tools

In order to select the type of the survey tool, the following parameters are typically considered:

12. Hole size/diameter and inclination angle,
13. Core barrel diameter,
14. Smallest drift in drill string,
15. Formation properties,
16. Temperature,
17. Mud properties.

   Generally, there are three available types of survey tools: Magnetic Film System (MFS), Electromagnetic Survey (EMS) tool (Fig. 10.10a) and Modular-Magnetic Tool (MMT) (Fig. 10.10b).

1. *Magnetic Film System (MFS)*:
   This method is old-fashioned and has the following drawbacks due to:

   1.1   Risk of stuck pipe in the hole due to the necessity of stopping the outer tube rotation and mud circulation, while the orientation pictures are taken.
   1.2   Additional rig time required due to the time required for the survey pictures to be taken (directly) and the time wasted due to lower ROP (indirectly).
   1.3   Risk of core breakage due to the torque of core bit after each picture is taken and thus higher risk of core jamming in the inner tube.

2. *EMS (Electromagnetic Survey[4])*:
   This tool has been illustrated in Fig. 10.10a, with its specifications as in Table 10.2.

   2.1   EMS makes continuous measurements using continuously scribing knives and the survey tool during the coring operations (unlike MFS method). That is coring does not have to be stopped to take the survey. Therefore, the

---

[4]In some sources, it stands for 'Electronic Multi-Shot Survey'.

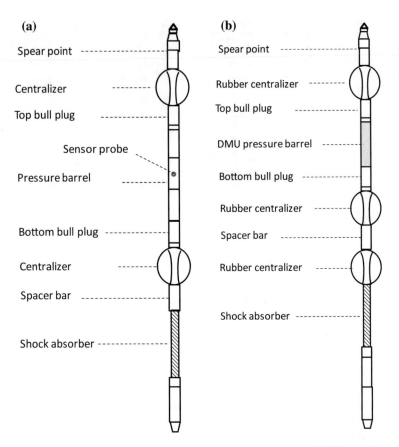

(a)

Spear point

Centralizer

Top bull plug

Sensor probe

Pressure barrel

Bottom bull plug

Centralizer

Spacer bar

Shock absorber

(b)

Spear point

Rubber centralizer

Top bull plug

DMU pressure barrel

Bottom bull plug

Rubber centralizer

Spacer bar

Rubber centralizer

Shock absorber

Fig. 10.10 Schematic of the survey tools; **a** electromagnetic survey tool, EMS, and **b** modular-magnetic tool, MMT (modified from courtesy of Baker Hughes GE)

possibility of core damage, scribe mark spiraling, core jamming, core barrel stuck in the hole, etc. are mitigated.

2.2   Using non-magnetic outer tubes and centralizers, the survey tool and the measured data are protected from magnetic interference.

2.3   At the end of coring operations and retrieval of the survey tool, the downhole data can be downloaded from the EMS memory/battery and a final tool face report is provided including the tool face orientation versus depth along with the hole survey data at the beginning and the end of the cored interval.

3. *The Modular-Magnetic Tool (MMT)*:

This tool has been illustrated in Fig. 10.10b, with its specifications as in Table 10.2.

Table 10.2 EMS and MMT specifications (published courtesy of Baker Hughes GE)

Parameter	EMS	MMT
Nominal D (in.)	1¾ (2 with heat shield)	1 3/8–1 3/4
Length (m)	3–6 ft	3–6
Temperature (°C)	125 (260 with heat shield)	150 (285 with heat shield)
Cycle per second	64	–
Tool capacity to save (no. of shots)	1023	
Max delay time (h)	10–36	200
Accelerometers number	3 (to measure inclination angle θ)	–
Magnetometers number	3 (to measure hole azimuth, AZ)	–
Min shot interval (s)	10	10
Measurement resolution (°)	0.01–0.1	0.01–0.1
Battery type and life (h)	Alkali (150)	Alkali (150) Lithium (300)

Baker Hughes GE, do not provide the survey instruments or survey services. The data below are generic information on the EMS and MMT tools provided earlier for these services. Baker Hughes GE still delivers the interface for connecting these tools provided by a survey vendor to the core barrels

3.1 It is a more advanced survey tool than EMS as it can be applied as a wireline steering survey tool easily. It can also provide real-time measurements through the wireline to transmit data to a surface computer equipped with powerful software. At the end of coring, it can be retrieved via wireline after shearing off an aluminum pin in the latch. In addition, it is equipped with an electronic memory single-shot or multi-shot survey tool.

3.2 Because of the special advanced features of the tool, its particular applications are highly deviated or horizontal wells, UBD, or air drilling, deep coring (due to high delay time up to 200 h), high temperature/geothermal wells and in short radius wells (radius of curvature < 11 m).

3.3 Indeed, positive latch must be used with MMT, which is a more advanced standard coring mule shoe: *Positive latch* is a modified mule shoe assembly. It anchors the survey tool to the inner tube (as a mule shoe).

Note: Currently, this tool is not typically used, except in very special cases.

References

Brindley, C.P. 1988. *Continuous Orientation Measurement Systems Minimize Drilling Risk During Coring Operations* SPE 17217-MS. Presented at the SPE/IADC Drilling Conference, 28 February–2 March, Dallas, Texas.

Brindley, C.P. and Sperry-Sun, N.L. 1988. *Continuous Orientation Measurement Systems Minimize Drilling Risk During Coring Operations*, SPE-17217-MS, Presented at SPE/IADC Drilling Conference, 28 February-2 March, Dallas, Texas.

Dennis, B., E. Standen, D.T. Georgi, and G.O. Callow. 1987. *Fracture Identification and Productivity Predictions in a Carbonate Reef Complex*, SPE 16808-MS. Presented at the 62nd SPE Annual Technical Conference and Exhibition, Dallas, Texas.

Hay, W., I. Knott, and G. Schiells. 1990. *Development of Coring and Electronic Data Logging Systems Enhances Data Quality and Quantity Facilitating Improved Oriented Core Analysis.* Presented at the SCA Conference in Dallas, Texas, USA.

Laubach, S.E., and E. Doherty. 1999. *Oriented Drilled Sidewall Cores for Natural Fracture Evaluation*, SPE 56801-MS. Presented at the SPE Annual Technical Conference and Exhibition, 3–6 October, Houston, Texas.

Nelson, R.A., L.C. Lenox, and B.J. Ward. 1987. *Oriented Core: Its Use, Error, and Uncertainty*, AAPG Bulletin, April 1987, pp. 357–367.

Rourke, M., and J. Torne. 2011. *A New Wireline Rotary Coring Tool: Development Overview and Experience from the Middle East*, SPE 149128-MS. Presented at the SPE/DGS Conference and Exhibition, 15–18 May, Al-Khobar, Saudi Arabia.

Skopec, R.A., M.M. Mann, D. Jeffers, and S.P. Grier. 1992. *Horizontal Core Acquisition and Orientation for Formation Evaluation*, SPE 20418-PA. Published in Journal of SPE Drilling Engineering, March, p. 47–54.

Chapter 11
Pressure/In Situ Coring

11.1 Introduction

It was already discussed in Chaps. 7 and 8 that using conventional, non-pressurized coring during tripping, the core may undergo some invasion-related mechanical damage due to the fluid expulsion. This causes some loss of data and inability to recover meaningful core analysis data such as the in situ fluid saturations (inferred from Johns and Lewis 1981; Hyland 1983; Bjorum 2013; Bjorum and Sinclair 2013; Ali et al. 2014; Cerri et al. 2015; Ashena 2017). The capture and characterization of the fluids being expelled out of the core sample during its tripping is a possible solution as it can provide information about the hydrocarbon volume and its properties. To address this issue, pressure/in situ coring has been already introduced in the industry. In this system (combined either with the conventional or wireline operations), at the end of coring, the inner tube assembly containing the core barrel is raised in a closed system to the rig floor, i.e., under its bottom-hole pressure.

Pressure coring is not a new concept as it was first proposed in the 1930s, but it remained a laboratory tool until the 1970s (Johns and Lewis 1981; Hyland 1983). This system was practiced until the end of the 1990s when some unsafe incidents due to some careless operations gave a wrong impression of the system. Recently, as high-cost projects and increasingly more stringent industry economics demand more accurate data for decision-making, the pseudo-pressure coring has again become very popular as a reservoir engineering evaluation method.

Therefore, in this chapter, following a discussion on the original and pseudo-pressure coring methods, the advantages and disadvantages of the system, and finally its operating procedures are covered.

© Springer International Publishing AG, part of Springer Nature 2018
R. Ashena and G. Thonhauser, *Coring Methods and Systems*,
https://doi.org/10.1007/978-3-319-77733-7_11

11.2   Original-Pressure Coring

In original pressure coring, the core specimen is retrieved to the surface in a completely sealed manner, by maintaining it at bottom-hole pressure (Sattler et al. 1988). When the core reaches the surface, in case the lithology is not shale, it may be frozen to immobilize the fluids and gases within the core (Johns and Lewis 1981; Hyland 1983). Then, it is sent to the core analysis laboratory, where its pressure is bled-off from the bottomhole to the atmospheric pressure very slowly. Finally, it can be used for core analysis.

Application of this original pressure coring system involves some issues. First, it has safety concerns (inferred from Hyland 1983; Ashena et al. 2016a, b) as it is required to handle extreme pressures even over 10 thousand psi, which is a tremendous pressure for handling and processing of the high-pressure core barrel at the surface. Second, technically, as the pressure coring tools have to stand high working pressures in traditional coring systems, considerable thickness is required for the tool. Therefore, the core diameter was rather small, confined to maximum 1–1.5-in. diameter, for proper core analysis (inferred from Hyland 1983; Ashena et al. 2016a). This limits the results obtainable from the core analysis. Considering these issues, particularly some unsafe practices and incidents, these tools are not yet in the market.

Although this original-pressure coring method is dated, this method has been recently translated to sidewall coring method with a pressure rating of maximum 25,000 psi (Pinkett and Westacott 2016). It has been recently field-tested and the interest for it shows an increasing growth because it can contribute to decision-making in terms of hole completion after the well has been drilled. As an example of using original-pressure coring concept, in 2014, Halliburton introduced *CoreVault* by updating their rather dated HTHP coring tool (which was *Hostile Rotary SideWall Coring Tool, HRSCT*) (Fig. 11.1).

11.3   Pseudo-Pressure Coring

Pseudo-pressure coring method and its tool design have been designed and proposed as an alternative to the traditional original pressure method mainly for safety reasons and obtaining larger cores. Using the pseudo-pressure coring system, the sample is brought up to the surface (or tripped out) in a closed canister. In case the containing pressure exceeds a specified value, e.g., 1000 psi due to gas expansion, the canister can be opened to release the excess pressure (Fig. 11.4). Therefore, compared with the original pressure coring system, the main differences are that it is equipped with pressure canisters, pressure relief valve, pressure and temperature transducers along the core barrel and canister(s), full-closure catcher system, and triple tube system (i.e., using a liner (e.g., aluminum) inside the inner tube, inferred

Fig. 11.1 A typical
traditional pressure coring
system (published courtesy of
Baker Hughes GE)

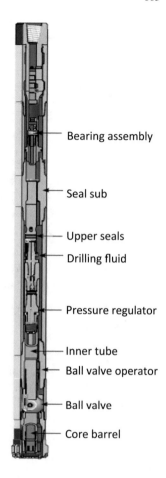

Bearing assembly

Seal sub

Upper seals

Drilling fluid

Pressure regulator

Inner tube

Ball valve operator

Ball valve

Core barrel

from Anis 2001). Because of the opening mechanism of the canister in case of extreme pressures, *pseudo* is attached to this pressure system.

Some considerations of the tool are as follows:

1. After core was drilled/cut and before tripping to the surface, the core barrel is closed and sealed to retain all the released gases and fluids.
2. Then, gas and fluids inside the core sample are allowed to expand while tripping using variable volume system or canister(s). Therefore, the gas and fluids are allowed to move from inner tube to the canister.
3. A single canister (previously two or three canisters) is positioned in BHA, directly above the inner tube so that any expelled or released gases or liquids can be captured there (inferred from Bjorum 2013; Schultheiss et al. 2010; Bjorum and Sinclair 2013).
4. Pressure and temperature transducers have been distributed throughout the top of the inner tube and also the canisters to measure pressure and temperature to monitor the operations and ensure the reliability or validity of the operations.

Table 11.1 Pseudo-pressure coring parameters (published courtesy of *Reservoir Group*)

Core D (in.)	3–4
Core L (in.)	10
Max pressure (psi)	1000
Max temperature (°F)	250 (125 °C)
Pressure/temperature transducers	Distributed throughout the barrel and canisters

5. At the surface, following the measurement of the canister gas volume at the surface, the pressure is bled-off in a controlled manner.
6. The coring tool is designed to run both on conventional drill pipe and also wireline (wireline continuous coring).

Some important parameters of the tool and the sample are given as follows in Table 11.1.

Since the sample is depressurized by storing the released gas during tripping, it provides the following benefits compared to the original pressure coring:

1. The maximum working pressure requirement of the coring tool has been reduced (Bjorum 2013; Bjorum and Sinclair 2013; Davis et al. 2013, etc.). Therefore, there is less risk of working with the system.
2. A larger core sample (diameter and length) can be obtained (even up to with 3½–4 in. in 8½-in. hole size) using *specialty-drill pipes* (5½-in. OD and 4.625-in. ID) and 4¼-in bore *jars* (Farese et al. 2013a, b), which contributes to more reliable:

   a. Estimations of the fluid volumes to be obtained.
   b. Well reservoir and PVT studies on the recovered gas or fluid samples.
   c. Gas composition.

The schematics of wireline pressure coring assemblies are given in Fig. 11.2 for the case of during coring, and in Fig. 11.3 for the case of during tripping on the way to the surface.

11.4   Advantages and Disadvantages

Although pseudo-pressure coring has been shown for its advantages to be potentially able to revolutionize the coring operations and enhance the quality of cores retrieved, it suffers from some issues. The advantages and disadvantages of this system are as follows:

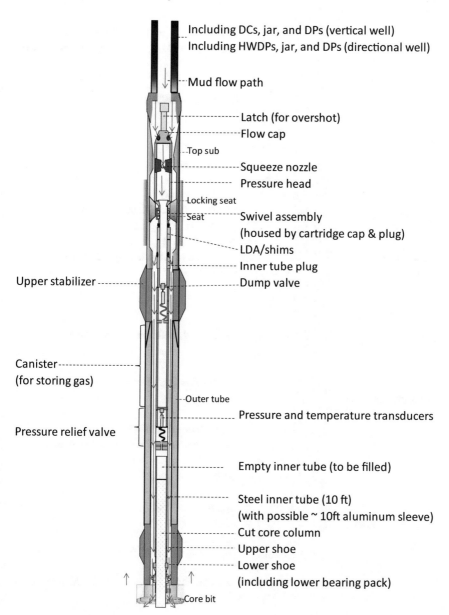

Fig. 11.2 Schematic of pseudo-pressure/in situ coring (during coring). The red arrows show the mud circulation flow path during coring

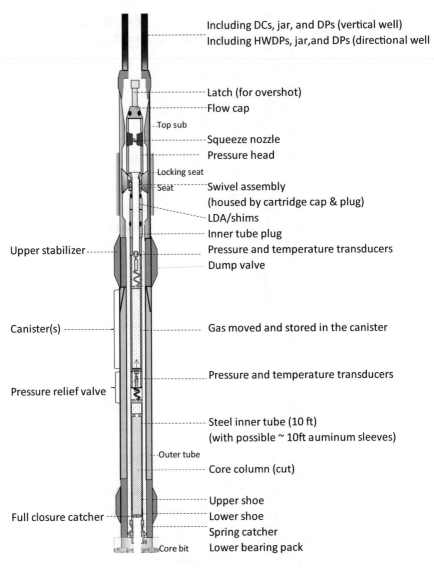

Fig. 11.3 Schematic of pseudo-pressure/in situ coring (after coring and during tripping, on the way to the surface). The red arrows show the flow path of the gas

Advantages

Compared with other coring systems, this system has several advantages as follows (inferred from Schultheiss et al. 2010; Bjorum 2013; Bjorum and Sinclair 2013; Al Neaimi et al. 2014; Ali et al. 2014; Cerri et al. 2015; Ashena et al. 2016a, b):

1. Reservoir fluid studies such as PVT, etc. using the captured gas and liquids in the canisters for recombination.
2. Accurate measurement of the initial hydrocarbon volumes in place.
3. Accurate gas and liquid composition (by gas chromatography).
4. Better estimation of the geomechanical properties particularly in tight cores as they undergo less damage during slow depreciation at the surface.
5. The flexibility of using with other systems:

   a. It is applicable/combinable with the wireline continuous coring system in addition to the conventional system (Yamamoto et al. 2014; Ali et al. 2014; Ashena et al. 2016a).
   b. It is possible to utilize special inner tube systems such as triple tubes (i.e., a non-split aluminum liner which is used inside the inner tube) for enhanced core protection. If a split liner is used (called *half-moon liner* by *Reservoir Group*), greater core recovery can be obtained.

Disadvantages
Despite the interesting advantages and its popularity during the recent years, it suffers from some drawbacks as follows (inferred from Al Neaimi et al. 2014; Ashena et al. 2016a) (Fig. 11.4):

1. Safety concerns still exist as the core barrel is retrieved with pressure (with maximum 1000 psi) to the surface.
2. No or little understanding of core fluid or gas flow and its volume with time exists while retrieving the sample to the surface. Still, some information about individual fluid phases (gas and oil at downhole conditions) is not provided as it is not clear when and at what depth the fluids are combined.
3. Only rather short cores (about 10 ft) can be taken due to the gas which should be captured and the maximum pressure restriction.
4. The sealing system used is complex and thus there is some chance of pressure seal failure, e.g., due to a leakage.

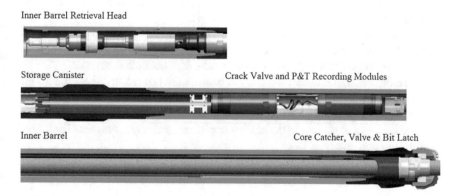

Fig. 11.4 A schematic of pseudo-pressure coring (Bjorum 2013)

5. There are additional costs due to very special core and rig site handling, lab analysis, and highly skilled personnel, which are required.
6. Finally, pressure coring is currently a very costly coring technique.

11.5   Operating Procedure

The operating procedures of the pressure coring resemble the original method, except for its surface handling (5.5). The tool operating procedures are as follows:

1. Check the BHA (i.e., the core barrel and the assemblies: the inner and outer tubes, jar, stabilizers, swivel, float, etc.) carefully before running in hole. *Large-bore jars* with ID drift of 4¼-in. is recommended to ensure the inner tube assembly can be successfully pulled through it. The OD of the core barrel (stabilizers) should not be smaller than 4 mm from the core bit OD.
2. Drill conventionally until the core point is reached.
3. Pull the drill string and run the coring tools into the hole.
4. Cut a core with the standard length (e.g., 10 ft).
5. Break the core at the bottom (by applying overpull on the coring assembly and make it off-bottom).
6. (For wireline pressure coring): Run down the wireline and overshot and latch onto the coring assembly.
7. Pull the coring assembly via two overpulls (The first pull allows the core to pass the full-closure valve assembly and seal the tool, and the second one separates the inner tube, which enables the tripping to commence).
8. (For wireline pressure coring): Adjust the tripping rate via wireline.
9. Recover the inner tube and canisters at the surface at the end of POOH.
10. Read the initial pressure and temperature readings. Download all the initial pressure and temperature measurement curves of the inner tube and canisters.
11. Separate the canister, heat only the canister to make the temperature reach the bottom-hole temperature to dissolve any hydrocarbon solids and gasify the hydrocarbon liquids (Al Neaimi et al. 2014).
12. Exit the gas through a flow meter and thus bleed-off the pressure to the atmosphere in a controlled manner (Al Neaimi et al. 2014).
13. Similarly degas/bleed-off the gas inner tube in a controlled manner.
14. Once degassed, the inner tube and the canister are opened to collect the liquids in graduated cylinders (Al Neaimi et al. 2014).
15. For tight gas-bearing cores (e.g., from shale or Coalbed methane), extract the core sample and place it in desorption canisters for enough duration. Then, send it to the core analysis lab.

References

Ali, M., G.H. Hegazy, M.N. Aftab, A.M. Negm, A.A. Syed, and A.H. Anis. 2014. *First Wireline and Elevated Pressure Coring in UAE—Saved 30% of Coring Time for Shallow Reservoirs & Delivered Realistic Fluids and Gas Saturations,* SPE 171866 MS. Presented at the Abu Dhabi International Petroleum Exhibition and Conference, November 10–13.

Al Neaimi, M.A., A.S. Tee, D. Boyd, R. Al Shehhi, E.A. Mohamed, K.M.N. Namboodiri, H.A. Junaibi, M. Al Zaabi, H. Al Braik, S. Ftes, A.K. Medjiah, A. Farouz, B. Gao, M. Gay, S.M. Tariq, B. Fudge, P. Collett, T. Gill, A. Abdul-hamid, B. Schipper, and Sheldon, M. 2014. *Acquisition of an Elevated Pressure Core in a Gas Flooded Carbonate Oil Reservoir: Design and Operational Challenges,* SPE 171815 MS. Presented at the Abu Dhabi International Petroleum Exhibition and Conference, November 10–13.

Anis. 2001. *Coring in the New Millennium,* SCA 2001-40. Presented at the SCA Conference, Edinburgh, Scotland, September 16–19.

Ashena, R., W. Vortisch, M. Prohaska, and G. Thonhauser. 2016a. *Innovative Concepts in Wireline Continuous Coring,* SPE 180017-MS. Presented at the SPE Bergen One-Day Seminar, April 20, Norway.

Ashena, R., W. Vortisch, M. Prohaska, and G. Thonhauser. 2016b. *Innovative Concepts in Wireline Continuous Coring,* SPE 0816-0060-JPT. Published in SPE Journal of Petroleum Technology, August, pp. 60–61.

Ashena, R. 2017. *Optimization of Core Tripping Using a Thermoporoelastic Approach,* Ph.D. Dissertation Submitted to Montanuniversität Leoben.

Bjorum, M. 2013. *A New Coring Technology to Quantify Hydrocarbon Content and Saturation,* SPE 167228 MS. Presented at the Unconventional Resources Conference, Calgary, Alberta, Canada, November 5–7.

Bjorum, M., and J. Sinclair. 2013. *Pressure Coring, a New Tool for Unconventional Oil and Gas Characterization,* GeoConvention 2013, Calgary, 6–12 May.

Cerri, R., Martini, S.D., P. Balossino, L. Gioacchini, I. Colombo, E. Spelta, M. Bartosek, and M. Bjorum. 2015. *Combined Application of Pressure Coring and Desorption Analysis for Barnett Shale Gas Evaluation,* SPE 172936. Presented at the Middle East Unconventional Resources Conference and Exhibition, January 26–28.

Davis, M., R. Williams, D. Willberg, M. Bjorum, D.M. Willberg, and K. Akbarzadeh. 2013. *Novel Controlled Pressure Coring and Laboratory Methodologies Enable Quantitative Determination of Resource-in-Place and PVT Behavior of the Duvernay Shale,* SPE 167199. Presented at the Unconventional Resources Conference, Calgary, Alberta, Canada, November 5–7.

Farese, T.M., A.K. Mohanna, H. Ahmed, I.A. Adebiyi, and Omar, A.A.F. 2013a. *Coring Optimization: Wireline Recovery Using Standard Drill Pipe,* SPE 166739. Presented at the Middle East Drilling Technology Conference and Exhibition, October 7–9.

Farese, T., H. Ahmed, and A. Mohanna. 2013b. *A New Standard in Wireline Coring: Recovering Larger Diameter Wireline Core Through Standard Drill Pipe and Custom Large Bore Jar,* SPE 163507. Presented at the SPE/IADC Drilling Conference and Exhibition, Amsterdam, The Netherlands, March 5–7.

Hyland, C.R. 1983. *Pressure Coring-An Oilfield Tool,* SPE 12093. Presented at the SPE Annual Technical Conference and Exhibition, San Francisco, California, US, October 5–8.

Johns, S.B., and D.J. Lewis. 1981. *Improved Pressure Coring in Unconsolidated Sands,* SPWLA 1981-U. Presented at the SPWLA 22nd Annual Logging Symposium, Mexico City, Mexico, June 23–26.

Pinkett, J., and D. Westacott. 2016 *Innovative SideWall Pressure Coring Technology Improves reservoir Insight in Multiple Applications,* 2016-G SPWLA. Published in the SPWLA 57th Logging Symposium, June 25–29.

Sattler, A.R., A.A. Heckes, and J.A. Clark. 1988. *Pressure Core Measurements in Tight sandstone Lenses during the Multiwell Experiment*, SPE 12853-PA. Published in the SPE Formation Evaluation Journal, 645–650.

Schultheiss, P., J.T. Aumann, and G.D. Humphrey. 2010. *Special Session-Gas Hydrates: Pressure Coring and Pressure Core Analysis Developments for the Upcoming Gulf of Mexico Joint Industry Project Coring Expedition*, OTC 20827 MS. Presented at the Offshore Technology Conference, Houston, Texas, May 3–6.

Yamamoto, K., N. Inada, S. Kubo, T. Fujii, K. Suzuki, Y. Nakatsuka, T. Ikawa, M. Seki, Y. Konno, J. Yoneda, J. Nagao, and Y. Mizuguchi. 2014. *A Pressure Coring Operation and On-board Analyses of Methane Hydrate-bearing Samples*, OTC-25305. Presented at the Offshore Technology Conference, Houston, Texas, May 5–8.

Chapter 12
Logging-While-Coring

12.1 Introduction

One of the main challenges during coring, particularly in exploration wells, is the possibility of unidentification of the right coring point/depth. Typically, this can be noticed only at the surface using gamma-ray logging or geology study. In such cases, the retrieved core sample is already obtained from the undesired formation or interval and thus, the success of the operations has been seriously challenged as a lot of money and efforts have been wasted. Facing this challenge, coring may be repeated (this time trying to be in the right depth interval). Ignoring to core and just relying on subsequent wireline logs for formation evaluation is not an option in an exploration well (because the logs would remain uncalibrated without core data and of limited value). Another challenge of coring (and generally formation evaluation) is that coring and wireline logging are taken under different times and thus well and formation conditions causing some adverse mud invasion or mechanical changes, and even depth matching issue (refer to Chap. 8).

To overcome these above challenges, Logging-While-Coring (LWC) system is proposed, which enables simultaneous downhole well logging and coring. First of all, it provides knowledge about the formation properties which is to be or being cored (Goldberg et al. 2003, 2004, 2006). Therefore, the identification of the appropriate coring depth interval (i.e., the initial and the final depth) is viable, which contributes to the proper determination of the depth to start and end the coring operation. Second, the log data found by LWC is more reliable for formation evaluation than subsequent wireline logging as it is taken before any adverse effects (such as mud invasion) have occurred to the formation. In addition, during LWC, the log data is taken simultaneously and at the same depth and under the same conditions with coring. Therefore, meaningful matching of the two main formation

© Springer International Publishing AG, part of Springer Nature 2018   171
R. Ashena and G. Thonhauser, *Coring Methods and Systems*,
https://doi.org/10.1007/978-3-319-77733-7_12

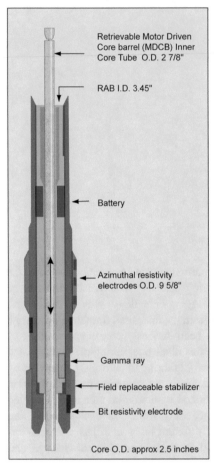 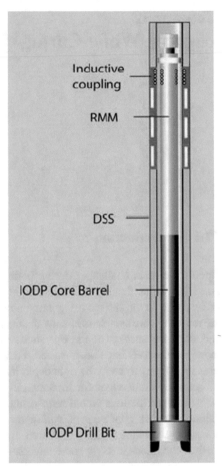

Retrievable Motor Driven
Core barrel (MDCB) Inner
Core Tube O.D. 2 7/8"

RAB I.D. 3.45"

Battery

Azimuthal resistivity
electrodes O.D. 9 5/8"

Gamma ray

Field replaceable stabilizer

Bit resistivity electrode

Core O.D. approx 2.5 inches

Inductive
coupling

RMM

DSS

IODP Core Barrel

IODP Drill Bit

(Goldberg et al. 2003 & 2004)   **(Myers et al. 2006)**

evaluation data, i.e., core and log data versus depth, is achievable, which contributes to better reservoir characterization including its structure and well completion decisions (inferred from Goldberg et al. 2003, 2004, 2006). Third, using
LWC, the necessity of wireline logging subsequent to the coring operation is
eliminated, which can compensate the additional cost caused by using LWC system. Fourth, using the specified LWC sensors, the drilling dynamics parameters can
be potentially measured at the core bit (Myers et al. 2006), which contributes to
coring optimization and obtaining high core recovery and quality (particularly for
the next coring jobs).

It is noted that there was a dated system of logging, while wireline continuous
coring (only for the drilling mode), in order to take the directional survey and just the
gamma-ray among the formation properties (refer to Fig. 6.5). Therefore, there were
no LWC systems until 2002 when the method was first field-tested by the Integrated
Ocean Drilling Program (IODP), as a pioneer in developing and testing LWC in

cooperation with some research centers. Then, one year later it was patented. For further information on the patents on LWC, refer to Chap. 15, Sect. 15.3.

Therefore, in this chapter, first, different downhole logging methods and systems are described. Next, the methods of the telemetry of the measured data from the bottom-hole to the surface are covered. Finally, the main challenge of the LWC system is discussed.

12.2 Methods of Downhole Logging

For downhole logging of the core interval of the formation, as shown in Table 12.1, there are three main options: (1) wireline well logging subsequent to coring, (2) installing Logging-While-Drilling (LWD) tools above the core barrel in order to

Table 12.1 Different methods of downhole logging of the core interval of the formation

Method	Measured parameters		Where	When	Schematic
Conventional open-hole wireline (WL) logging	− Natural gamma-ray (GR) − Caliper − Porosity logs (neutron, density, sonic[a]) − Electrical resistivity − Dipole sonic[b]		− Performed after the termination of coring	After coring	
LWD tool	− Natural Gamma Ray − Electrical resistivity − Neutron − Bulk density − Annular Pressure While Drilling (APWD)	By Ecoscope[c]	− Installed above the core barrel	While coring	
	− MWD telemetry	By TeleScope			
	− Fracture interpretation − Breakout interpretation	By GeoVISION			
	− Dipole sonic	By SonicVISION			

(continued)

Table 12.1 (continued)

Method	Measured parameters	Where	When	Schematic
LWC	− Natural gamma-ray − Resistivity (measurement of other parameters is possible if the tools are modified). − Density − Temperature − In situ pressure	− Installed inside the core barrel (on the outer tubes or between the inner and outer tubes)	While coring	

[a]Usually only the compressional wave velocity and the transit time is measured

[b]If dipole Sonic Imager (DSI) is utilized, the compressional wave velocity (V_p) and shear wave velocity (V_s) can be measured with depth. This is valuable for wellbore stability and rock mechanical properties

[c]The Annular pressure while drilling measurement contributes to detection of any annular pressure decrease and possible flow/kick into the wellbore. For instance, a decrease in the annular pressure of, e.g., 100 psi can alarm the emergency

log the formation for about 20–30 m above the bit (Lovell et al. 1995), and (3) installing LWC tools in the core barrel, on the outer tube or between the inner and outer tubes (Goldberg et al. 2003, 2004, 2006). The comparison of these methods is given in Table 12.2.

It is inferred from Tables 12.1 and 12.2 that the wireline logging does not represent the real pre-drilled properties of the reservoir formation and require some extra rig time although it can measure a wide range of parameters such as the gamma-ray, resistivity, neutron density, temperature, stress orientation, etc. For the LWD method, the range of data that can be obtained is as broad as the wireline (and even wider); however, they can be only measured about 30 m above the core barrel (Lovell et al. 1995), which limits its value. The same data can be theoretically measured in the LWC method with this advantage that they are measured almost at the bit.

The methods discussed above correspond to downhole logging of the formation core. It is also noted that the recovered core samples at the surface are typically logged by a gamma-ray sensor (refer to Chap. 14 for more information), which is used just for depth matching purposes or deciding on further coring. Therefore, due to its limited contribution, it cannot really replace the downhole core logging method.

Table 12.2 Comparison of different methods of downhole logging of the core interval (inferred from Goldberg et al. 2003, 2004, 2006)

Method	Practice	Advantages	Disadvantages
WL logging	Routine practice	More widespread and in practice	– More rig time due to necessity of tripping – Invasion incurred measurements – Stability of the formations after coring
LWD tool	Successfully applied by some IODP expeditions, Baker Hughes, etc.	– Can provide real-time measurements along with coring – Mitigated wellbore instability challenge and well control – A larger number of geophysical or petrophysical parameters measured (more than wireline logging and LWC)	– Measurements do not correspond to the bit (i.e., 30–40 m above the bit) and thus cannot help precise coring navigation – More expensive than WL logging
LWC	Successfully applied by some IODP expeditions	– Less expensive than LWD – As wireline is used to retrieve the core to the surface, no conventional drill pipe retrieval is required (time-saving) for retrieving the core barrel – Measurements are precisely matched with respect to the core depth (i.e., precise core log depth matching) – The can provides more reliable measurements than the wireline method and LWD (less invasion to the formation while logging) – Mitigated wellbore instability challenge and well control – The validity and credit of LWC logs are higher than LWD logs	– Could be more expensive than WL logging – Logging tools must be modified for LWC measurements – Still requires further development

12.3 LWC Systems

There are two field-tested LWC systems: the first system has been developed for measuring resistivity, gamma-ray, and annular pressure; the second one has been developed for measuring the drilling dynamic properties. In the first system, called *Resistivity At Bit (RAB)* (Fig. 12.1), the resistivity sensors equipped with memories, are mounted on the outer tube and the bit; the gamma-ray sensor and the battery are placed between the outer and inner tubes to charge the sensors (Goldberg et al. 2003, 2004, 2006). In the second system, called *Downhole Sensor Subs (DSS)* (Fig. 12.2), coring dynamic parameters are measured using the sensors. Then, it is either transferred to the surface in real-time manner or/and saved in the retrievable memory (RMM) to be retrieved each time the core reaches the surface via the wireline (Myers et al. 2006). These systems are further discussed below:

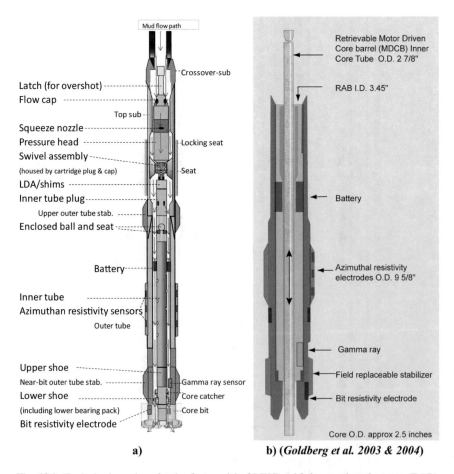

a)   b) (*Goldberg et al. 2003 & 2004*)

Fig. 12.1 Typical schematics of **a** the first model of LWC and **b** its core barrel system (RAB)

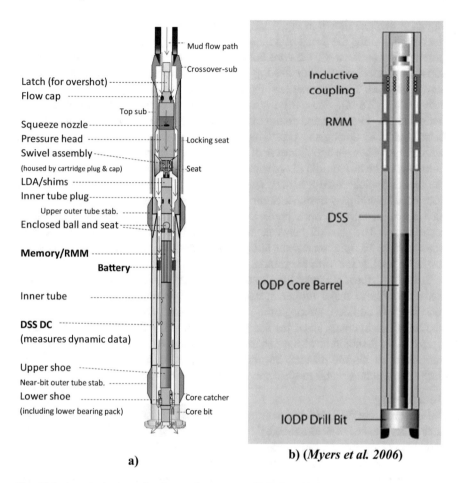

Fig. 12.2 A typical schematic of **a** another system of LWC and **b** its core barrel (DSS-RMM)

12.4 RAB System

Tool Description

In 2002, the first type of LWC system called RAB was examined and field-tested to measure the gamma-ray and resistivity during coring in leg-204 (off coast Oregon) by Ocean Drilling Program (ODP), which was exploring for gas hydrates (Trehu et al. 2003). As Lovell et al. (1995) and Goldberg et al. (2004) depicted, it was already developed by installing a resistivity tool just at the bit, a wireline-retrievable core barrel, and a latching tool (usually Motor Driven Core Barrel (MDCB)). Core samples with the size of about 2 ½-in. were obtained in 9 7/8-in. hole size. This was the first simultaneous use of coring and logging techniques in practice. The schematic of the tool is shown in Fig. 12.1.

In this system, as shown in Fig. 12.1, as described in Goldberg et al. (2003, 2004), the core barrel fits the throat of a modified Schlumberger's resistivity just at the bit (abbreviated as RAB-8), the core barrels used were non-magnetic in order not to cause any interference with the logging tools (in terms of the orientation measurement, etc.). This was because the coring tool was developed for a larger hole size (than 9 7/8-in). Thus, a new resistivity button sleeve and a stabilizer were manufactured to accommodate the bit size of 9 7/8-in. Only the motor-driven core barrel (MDCB) among the ODP's coring systems, was sufficiently narrow to fit within the 3.45-in. inside diameter of RAB-8. Minor modifications to MDCB core barrel system was required in order to adjust the length and latching mechanism of the tool. A typical RAB-8 battery was placed in the annulus such that the MDCD core barrel could pass through. Some electrodes were placed a little up the bit and also some stabilizers were predicted for measurement and trajectory control, respectively. The standoff between the modified RAB tool and the borehole wall was 0.185-in. No mud motor was utilized, but rather the core barrel corresponded to a similar application with the motor.

It is noted that by modifying the system and adding other sensors, it is potentially possible to make further measurements in addition to the gamma-ray and resistivity. In addition, although the LWC in this case study was used for shallow depths of gas hydrates under the sea floor, the same LWC tool can be utilized for deep coring. It should be also noted that the measured data were not retrieved to the surface in a real-time manner, but they were recorded in a memory and retrieved each time the inner tube reached the surface via wireline.

Tool Testing

After fabricating the LWC tool comprising MDCB and RAB-8 components, it was successfully tested at Schlumberger's Genesis Rig in Sugar Land, Texas in 2003 by properly coring through low-grade cement. Afterward, the LWC system was sent to Oregon coasts to be tested by ODP in real coring conditions (ODP legs, 204 and 209). After the test, the main logging component of LWC system (RAB-8 tool) was calibrated and it showed that the logs taken had been reliable (2003, 2004).

Logs and Interpretation

Figure 12.3 shows the three logs obtained by the RAB. The log tracks show the log data including the GR and resistivity; in addition to the data obtained from the core analysis including the density, magnetic susceptibility (for detection of the lithology and mineralogy of formations), and the coring ROP (2003, 2004). These measurements can be used for the identification of lithology of the formation in exploration drilling. For instance, it can be inferred from the log results that the formation lithology has changed at 60 m below the sea floor as the three logs show abrupt deviations, which can be indicative of dissociation of gas hydrates.

The same figure compares the logs obtained by LWC (from one well: 1249-A) and LWD (from another adjacent well: 1249-B). There is a difference between the GR responses of LWC and LWD. There are several reasons for this: (1) there was a

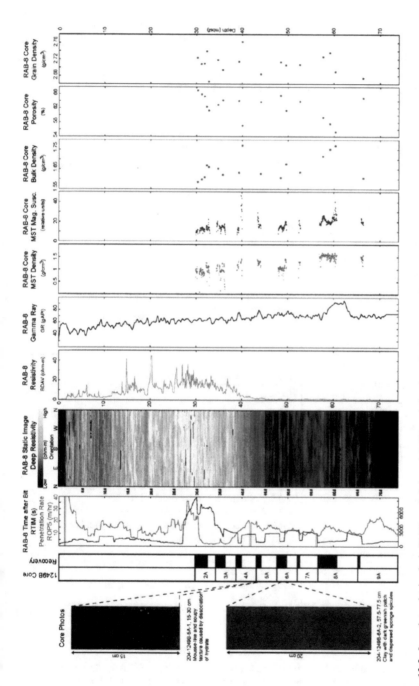

Fig. 12.3 Log data obtained via RAB, resistivity, and GR and the data obtained from the core analysis (Goldberg et al. 2003, 2004). Bulk density, grain density, and porosity were measured from the core analysis of discrete sample plugs

difference between the standoff of the LWC and LWD tools to the borehole wall (refer to Table 12.3), (2) the time after bit[1] in the LWC system was different from the LWD, and (3) There was 40 m lateral distance between the two wells, and 0.5 m difference in the water depth. Please note that the LWD tool (consisting of Schlumberger's *GVR-6/GeoVision resistivity tool*, and Schlumberger's *VDN tool*, sub for nuclear measurements) has been run to a greater depth in well 1249-A than the LWC (or RAB-8 tool) in well 1249-B (Fig. 12.4).

Table 12.3 Specifications and performance of the RAB tool (Goldberg et al. 2003, 2004)

Location	Offshore Oregon, crest of southern hydrate ridge, Region: leg 204, LWC
Geology and type of hydrocarbon explored	Shales, Gas hydrates (offshore)
Depth (m)	30–75
Seawater depth (m)	About 790
Cored interval (m, below sea floor)	45
Total no. of cores in all holes	8
No. of coring	2 (1st and 2nd length of core: 4.5 and 9 m)
No. of cores using plastic liners	2
Hole size or core bit size (in)	9 7/8 (This size is the standard ODP size)
Type of bit	4-cone
RAB-8 ID (in)	3.45
Core size (in)	2.56 (65 mm): standard cores 2.35 (60 mm): cores in plastic liners
$ROP_{ave.}$ (m/h)	= 8 As it was difficult to control WOB in the soft shallow sediments, it was attempted to keep ROP constant
Average core recovery (%)	32.9% (for cores taken by standard method) 42.3& (for cores taken in plastic liners) The average core recovery was low as the core barrel was designed for hard formations
Max. core recovery (%)	67.8%
LWC tool standoff from the hole wall (in)	= 0.185 (4.7 mm) In typical LWD tools, using GVR-6 tool, the standoff is greater (0.375-in.)

(continued)

---

[1]Time after bit is the time taken from the moment the bit cuts the well (at the deepest point of the coring string) until a measurement is made by a specific sensor (at a point higher up in the string). This time depends on the distance between the bit and the sensor, and the ROP.

Table 12.3 (continued)

Core barrel selection	Based on fixed RAB tool
Core barrel system compatible with RAB	Motor Driven Core Barrel (MDCB). This coring system is sufficiently narrow to fit within the annulus of RAB. However, some modifications of MDCB were required to accommodate for the RAB length and latching mechanism
RAB-8 battery	Placed in the annular space between the outer and inner tubes
Trajectory control	Stabilizers
Measurement of resistivity	Resistivity electrodes
Logs	High-quality logs (resistivity, magnetic susceptibility, GR, ROP, time at the bit, etc.) recorded in the downhole tool memory. Bulk and grain density and porosity are measured on discrete samples
Log data processing	Done post-cruise and correlated to LWD results in nearby wells
LWC capability	Proper for deployment in harder formations with high resistivity. The system has a deficiency to retrieve core samples from soft formations
Core processing and archiving	Onboard of the rig

12.4.1 DSS-RMM System

The second LWC system, a system using Downhole Sensor Subs (called DSS-RMM) has been developed to measure the drilling dynamic parameters while coring. This system consists of two main components: (1) the DSS drill collar and (2) the memory called Retrievable Memory Module (RMM) (illustrated in Fig. 12.2 and listed as the first two items in Table 12.4). The DSS drill collar is an instrumented non-magnetic drill collar developed to measure the dynamic parameters while coring (i.e., WOB, TOB, the annular pressure, and temperature) and also the acceleration measurements to measure the drill ship motion (Guerin and Goldberg 2002; Myers et al. 2006). It is linked to the second main component, i.e., the RMM, using the coils (shown in Fig. 12.5) so that the measured data from the DSS can be transferred to the RMM. Each time the inner tube (containing the rock sample) is retrieved at the surface via wireline, the RMM is received at the surface and the data is downloaded. It is noted that the corresponding electronics and battery are fastened on the wall of DSS drill collar. The design of the DSS-RMM system, with the list of the dimensions shown in Table 12.5, was developed into a model and was field-tested by IODP in 2003 for obtaining a 10 ft long, 2.5-in. diameter core sample in the 8 ½-in hole size (Myers et al. 2006; Goldberg et al. 2006). In Fig. 12.6, the data measured by DSS and the received data in RMM have been compared, which indicates a good link between DSS drill collar and the RMM.

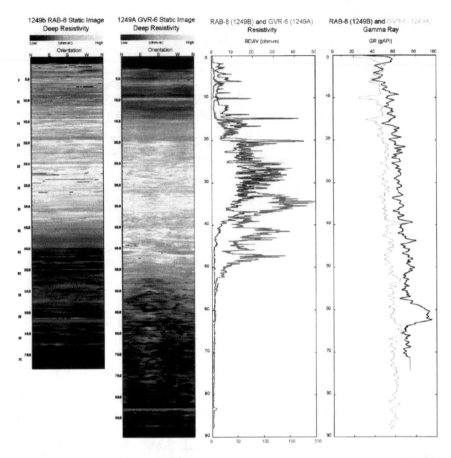

Fig. 12.4 Comparison of the measurements obtained by the LWC (using RAB-8 and MDCB tools) for a well (1249-A) and the LWD (using GVR-6 and VDN tools) for an adjacent well (1249-B). The depths are in meters (Goldberg et al. 2003, 2004)

12.5   Data Telemetry

The telemetry methods of the logging data fall into three categories of (1) only memory, (2) near real-time, and (3) the real-time.

Memory
In this method, the measured log data is measured and stored downhole. When the core barrel is conventionally pulled out of the hole, the data is then retrieved at the surface. Pulling the pipes out may be done after, e.g., 8 h to one day. Therefore, this method is only good to determine long-term trends of the drilling dynamics and cannot be used for optimization or quick decision-making.

Table 12.4 Main components of DSS-RMM system and their functions (data gathered from Myers et al. 2006; Goldberg et al. 2003, 2004, 2006)

Component	Developer	Location	Function
(1) DSS/ Instrumented DC	Texas A&M University	Above core head/bit	− Measures the WOB
			− Measures the Torque on Bit (TOB)
			− Measures the Annular bottom-hole pressure and temperature
(2) Retrievable Memory Module (RMM)	LDEO (Lamont Doherty Earth Observatory)	Top of the core barrel or the inner tube, e.g., APC (Advanced Piston Corer), RCB (Rotary Coring Bit), XCB	− Records the DSS data: pressure and the drilling dynamic data (including drill string dynamic parameters)
			− The recorded data can be retrieved at the end of coring when the core bit reached the surface
			− Measures the acceleration: accelerometers measure the ship motion to estimate the position of the ship relative to the bottomhole
Datalink System	LDEO		− Retrieves the data from the instrumented drill collar and inductively transfers it to the RMM of the core barrel/inner tube, then retrieves the RMM and the inner tube after coring every 9.5 m (Near Real-Time)

Fig. 12.5 Application of coils to transmit the data from DSS DC to the RMM core barrel (inductive data link system) (modified from Goldberg et al. 2006)

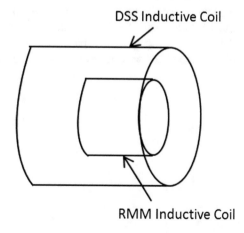

DSS Inductive Coil

RMM Inductive Coil

Table 12.5 The dimensions used in DSS-RMM system (data gathered from Goldberg et al. 2006)

Hole size (in.)	8 1/2
OD of DSS (in.)	8 1/4
ID of DSS (in.)	4 1/8
L of DSS (ft)	10
DSS wall thickness (in.)	5/16
ID of drill pipe	4 1/8
OD of core barrel (in.)	3 1/2
Core diameter (in.)	2.38
ID of core head/bit (in.)	3.8
Inner tube length (ft)	30 (range II pipe)
DSS type	Non-magnetic
Maximum DSS battery capacity (day)	4.5
Parameters measured by DSS	WOB, TOB, P_{ann}, P_{well}, etc.

Near Real-Time Telemetry

If the LWC system is used with wireline continuous coring method, retrieving the core barrel (i.e., inner tube) and the memory/RMM to the surface can be accomplished quickly via the wireline (inferred from Goldberg et al. 2003, 2004; Myers et al. 2006; *and LWC patents in* Sect. 15.3). In other words, e.g., following every 10 m of coring, the inner tube together with the memory of recorded measurements is retrieved to the surface; at the surface, the data in the memory can be downloaded. This data telemetry method is quite popular. As the tool development with time has been shown in Table 12.6, the design has moved to making near real-time logging. Compared with the memory method, this significantly increases the frequency of the data retrieved and contributes to better decision-making.

Real-Time Telemetry

The real-time data telemetry to the surface is practically possible if the MWD sub and commercial mud pulse systems are utilized. As shown in Fig. 12.7, the MWD sub is placed either above the core barrel or between the outer and inner tubes, but it is anyway connected to the LWC measuring sensors. The maximum frequency of the data recovery to the surface ranges from 0.5 to 1 Hz. In offshore wells, tuning of the data is necessary by comparing the heave response between the bit sensors and the up-hole. This telemetry method has not yet been field practiced.

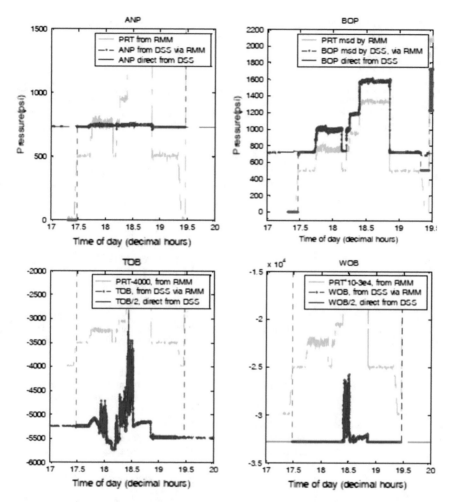

Fig. 12.6 The dark blue color (representing the DSS measurement) and the dark red (representing the data received in RMM) fall on each other, which is indicative of proper data transmission between the sensors (Myers et al. 2006; Goldberg et al. 2006)

Challenges

Practically, the LWC (i.e., RAB and DSS) systems are considered rather costly by the industry, even by the IODP Programs. Thus, currently, RAB and DSS (i.e., LWC) are not commonly practiced in coring. Thus, LWC technology has not been used since the early 2000s. However, to overcome this, the LWC devices are still under further development by IODP, Texas A&M University and the Borehole Research Group at LEO: Lamont Earth observatory as a research unit at Columbia University.

Table 12.6 The development process of DSS system (for near real-time measurement of the drilling dynamic data) (data gathered from Goldberg et al. 2003, 2004, 2006)

Year	Construction	Objective	Stage	Data recovery to the surface
1997	Instrumented drill collar by Drill String Acceleration (DSA)	Minimizing the negative impacts on the coring (in order to quantify the heave compensation required in offshore operations)	− It is activated before the deployment of the instrumented core barrel in the well − A modified logging tool (called DSA) is simply attached to the top of the core barrel − It stores the acceleration and pressure data in its memory − It measures the drill pipe motion	After the core barrel recovery (i.e., drill pipe tripping)
2000	Instrumented drill collar by DSS	Measuring more parameters	It measures the drilling dynamic data: WOB, TOB, annular pressure and temperature, etc.	After the core barrel recovery
2003	Instrumented drill collar and core barrel by DSS + RMM	Data transmission to the inner tube (near real-time) and then each time the inner tube is retrieved to the surface via wireline	− It measures and transfers the drilling dynamic data: WOB, TOB, annular pressure and temperature, etc. − Add an inductive coil on the DSS + one on the RMM core barrel + modified electronics and software − Frequency of measurements (one Hz, or every one second)	After the inner tube recovery

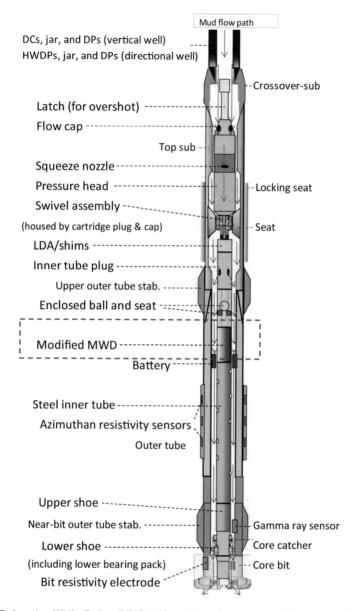

Fig. 12.7 Logging-While-Coring (LWC) with real-time data telemetry. Compare with Fig. 12.2a

References

Goldberg, D., G. Myers, K. Grigar, T. Pettigrew, S. Mrozewski, C. Arceneaux, and T. Collins. 2003. *Logging-While-Coring-New Technology Advances Scientific Drilling*, 2003-CC SPWLA. Presented at the SPWLA 44th Annual Logging Symposium, 22–25 June, Galveston, Texas.

Goldberg, D, G. Myers, G. Iturrino, K. Grigar, T. Pettigrew, S. Mrozewski, and ODP leg 209 Shipboard Scientific Party. 2004. *Logging While Coring-First Tests of a New Technology for Scientific Drilling*, 2004-V45N4A1 SPWLA. *Journal of Petrophysics* 45 (4): 328–334.

Goldberg, D., G. Myers, K. Iturrino, T. Grigar, T. Pettigrew, and S. Mrozewski. 2006. *Logging While Coring-New Technology for the Simultaneous Recovery of Downhole Cores and Geophysical Measurements*, from, *New Techniques in Sediment Core Analysis*. Handbook Edited by Rothwell, R.G., published by Geological Society of London, Special Publication 267.

Guerin, G., and D. Goldberg. 2002. *Heave Compensation and Formation Strength Evaluation from Downhole Acceleration Measurements While Coring*. Published in Springer Geo-Marine Letters, July, 133–141.

Lovell, J.R., R.A. Young, R.A. Rosthal, L. Buffington, and C.L. Arceneaux. 1995. *Structural Interpretation of Resistivity-At-the-Bit Images*, 1995-TT SPWLA. Presented at the SPWLA 36th Annual Logging Symposium, June 26–29.

Myers, G., D. Schroeder, W. Keogh, K. Grigar, and W. Masterson. 2006. *Coring Dynamics: Data Acquisition While Coring*, 17920-MS OTC. Presented at the Offshore Technology Conference, 1–4 May, Houston, Texas, USA.

Trehu, A.M., G. Bohrmann, F.R. Rack, and M.E. Torres. 2003. *Proceedings of the Ocean Drilling Program, Initial Reports*, Leg 204 Summary (online), Available in http://wwwodp.tamu.edu/publications/204_IR/204ir.htm (last accessed in July 2017).

Chapter 13
Other Coring Systems

13.1 Introduction

There are some other coring systems, which are not in widespread use, however, they are greatly important for particular applications. These coring systems include motor coring, underbalanced coring, and coil tubing among which motor coring has greater applicability in industry. These will be explained in this chapter.

13.2 Motor Coring

Directional and horizontal wells and very hard formations with induced fracture risk causing some challenges for coring. The directional and horizontal well design may give a lot of torque and drag to the coring string and create an environment downhole, where the parameters monitored at the surface do not reflect the actual coring parameters. Therefore, the coring string is not capable of providing a continuous and stable rotation transferred from the drill floor and it is also difficult to distribute the load over the cut area evenly. This causes lateral and torsional vibration, lower than enough rotation, high and uneven torque, and WOB which can result in inefficient coring and induced fractures. In hard rocks, there is a high risk of induced fractures due to non-optimized operational parameters. Induced fractures may lead to core jamming and unprecedented termination of the job or in the best case can lead to retrieval of low-quality core samples.

To overcome the above challenge, motor coring is proposed. In this method, the positive displacement motor (PDM) is connected just below the *top sub* and placed above the core barrel to change the trajectory whenever required. As motor coring is used for hard and abrasive formations, normally impregnated diamond or PDC bits are used in this system. A schematic of motor coring system is illustrated in Fig. 13.1. Operationally, it is performed with lower WOB, but instead with rather

© Springer International Publishing AG, part of Springer Nature 2018   189
R. Ashena and G. Thonhauser, *Coring Methods and Systems*,
https://doi.org/10.1007/978-3-319-77733-7_13

high RPM, and with smooth torque. The lower WOB contributes to lower possibility of induced formation fractures, higher enough RPM[1] is provided by the motor, and not by the coring string, (which compensates for the lower WOB), and the smooth operation enables rather even distribution of the rotation and load around the cut area. Therefore, motor coring contributes to reducing the BHA vibration and whirl, increasing its stability, while maintaining an acceptable ROP (*inferred from* Miller and Huey 1992). This mitigates the possibility of induced fractures ahead the bit and increases the core efficiency and recovery (Miller and Huey 1992; Rathmell et al. 1998). In case of necessity for greater BHA stability, e.g., with more than two outer tube pipes, it is recommended to use more than two stabilizers (as shown in Fig. 13.1), or to place a thruster on top of the PDM to create additional constant WOB.

It is noted that motor coring has the flexibility to be combined with other coring systems such as the wireline continuous and the invasion-mitigations system (Storms et al. 1991; Miller and Huey 1992; Rathmell et al. 1998; Fleckenstein and Eustes 2003; Deschamps et al. 2008[2]).

Enclosed Ball and Seat

In motor coring configuration, it is not possible to seal the inner tubes by dropping the ball from the surface. Therefore, it is a common field practice for motor coring that closed core barrel is run in the hole with the ball already in the seat. This would expose the barrel for the possibility of getting fill inside the inner tube and leading to an early jam. Therefore, instead a Downhole Activated Flow Diverter (DAFD) or drop-ball, sub is sometimes placed between the motor and the inner tube so that just before coring, the ball can be hydraulically activated by a circulation rate increase.

Horizontal Coring

In horizontal sections, usually two types of coring systems are applied: long radius (when the coring radius of curvature is greater than 200 m) and medium radius (when the coring radius of curvature is approximately between 80 and 200 m). The long-radius coring system usually uses conventional high-torque core barrels. However, the medium-radius system includes modified high-torque core barrels which are stabilized using special straight-blade stabilizers.

Combination Possibility

It is also possible to perform oriented motor coring by adding the scribe knives and the survey tool to the tool (for more information about oriented coring, refer to Chap. 10).

---

[1]Example 200 RPM. For making a comparison, it is noted that the RPM provided by the coring mud motor is lower than the drilling one because of lower mud flow rate in coring. The RPM should also match the bit design.

[2]http://www-odp.tamu.edu/publications/tnotes/tn31/mdcb/mdcb.htm. Accessed on July 10, 2017.

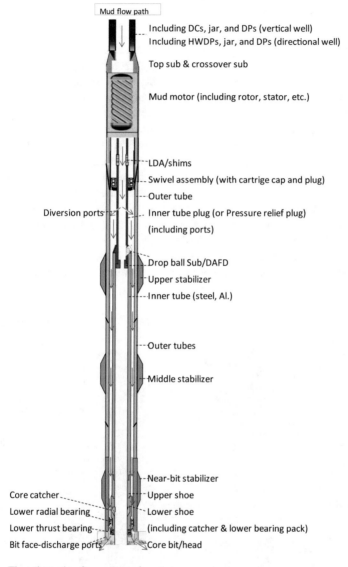

Mud flow path

Including DCs, jar, and DPs (vertical well)
Including HWDPs, jar, and DPs (directional well)

Top sub & crossover sub

Mud motor (including rotor, stator, etc.)

LDA/shims
Swivel assembly (with cartrige cap and plug)
Outer tube
Inner tube plug (or Pressure relief plug)
(including ports)

Diversion ports

Drop ball Sub/DAFD
Upper stabilizer
Inner tube (steel, Al.)

Outer tubes

Middle stabilizer

Near-bit stabilizer
Upper shoe
Core catcher
Lower radial bearing
Lower shoe
Lower thrust bearing
(including catcher & lower bearing pack)
Bit face-discharge ports
Core bit/head

Fig. 13.1 The schematic of a motor coring system

Due to the existence of the motor on top of the core barrel, normally the combination of motor coring with wireline continuous system is not viable as it is not possible to pull the core via wireline through the motor or measurement while drilling sub. Therefore, the motor coring system is used with the conventional (pipe) method.

13.3 Mini-Coring

Generally, most geomechanical studies are conducted in reservoir formations as most available cores have been taken from reservoirs only. However, for some reason such as investigation of wellbore stability issues in non-reservoir formations, geomechanical studies are necessary in such formations. In addition, some critical lithology detections such as casing point detection in non-reservoir formations may not be appropriately accomplished via cuttings merely. Therefore, coring in non-reservoir formations can be greatly beneficial, which can be done in a cost-effective manner using mini-coring In this method, a specific bit is used which generates micro-cores (with diameters ranging from 1 to 4 cm and 3 cm length) of formations during conventional drilling operations (Deschamps et al. 2008). This method has additional advantages such as increasing the drilling ROP. This system is the latest *Tercel*'s bit technology innovation, resulting from an extended R&D program led in collaboration with TOTAL S.A.

The bit design concept, which is shown in Figs. 13.2 and 13.3, is briefly explained as follows (Deschamps et al. 2008; Shinmoto et al. 2012):

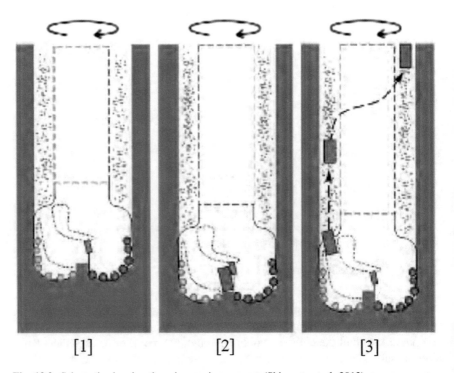

Fig. 13.2 Schematic showing the micro-coring concept (Shinmoto et al. 2012)

Fig. 13.3 An example of a 6-in micro-core bit (Deschamps et al. 2008)

- The core is generated in the center of the bit, where the cutting structure is interrupted.
- The micro-core, advancing to its maximum length, finally reaches a PDC stop which applies a lateral force at the top of the core, inducing a failure at its base.
- The core is led by the mud flow to the annulus via an evacuation area.

The key to successful core generation lies in the control of the combination of parameters affecting the overall bit/rock interaction and influencing the cutting mechanism (ductile or brittle mode), explained by (Deschamps et al. 2008): If a brittle cutting mode exists, characterized by the propagation of fracture in front of the cutters, huge cuttings chips will be generated. This phenomenon is amplified in the center area, leading to the shearing of the core at the earliest stage of its generation. In contrast ductile mode, characterized by plastic flow small cuttings ahead of the cutter face allows the preservation of the core in the center area. This ductile cutting mode must be promoted to enhance the core generation and therefore, recovery. Figure 13.4 shows an example of the difference in shapes and size of the cuttings resulting from ductile and brittle mode.

Fig. 13.4 **a** Ductile cuttings, **b** micro-cores, and **c** brittle cuttings (Deschamps et al. 2008)

13.4   Coiled-Tubing Coring

The developments of the coiled-tubing system particularly recently, have shown to revolutionize the drilling industry including the coring (Leising and Rike 1994). Coiled tubing reduces the tripping time and can thus greatly contribute to the economics of the job. Traditionally, coiled tubing was ideal for coring in slim-holes. Therefore, in a case study with the hole size of 4 1/8–4 3/4-in., using coiled tubing with the tube OD of 3½–3¾-in., cores with 1¾–2-in. diameter were obtained. Recently, the systems have been modified and some coiled-tubing systems for coring are available with the tube ODs of 4¾-in., 5¾-in., or 6 ¼-in. There have been some recent case studies for the application of this method. In a case study in the east of Texas, coiled tubing was successfully applied with a 2 7/8-in. mud motor to cut the number of two 3-m long, 2-in. diameter core samples at the depth of 4572 m from *Cotton Valley Reef*. In another case study, it was applied with a 4¾-in. motor in Western Canada to cut 8.3-m long cores from 6¼-in. and 7 7/8-in. hole sizes in the tar-sand reservoir. The cores were successfully cut with the average ROP of about 5.7 m/h and the recovery of 95%.

Although wireline continuous coring may be sometimes admired as the most cost-effective method particularly in exploration wells, the coiled-tubing system can potentially rival owing to its several advantages over the wireline method, which include:

- It can be combined with mud motors, which contributes to better trajectory control compared with wireline coring.
- It is also possible to obtain rock samples larger than 3½-in. diameters. There is no need for using the specialty-drill pipes as required by the wireline method.
- It is possible to circulate the mud through the inner tube while running in the hole to prevent fill or junk entry into the inner tube.

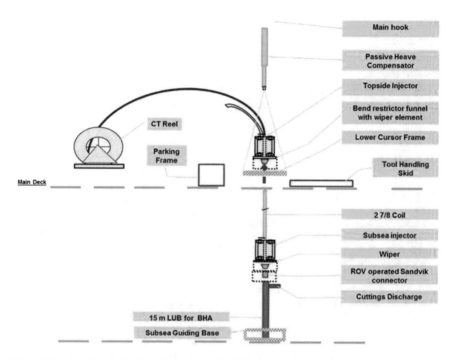

Fig. 13.5 Typical coiled-tubing rig for coring (published courtesy of Baker Hughes GE)

- It can be more effective than the wireline method particularly for medium deep applications.

It should be, however, reminded that coiled tubing still faces some limitations for its widespread use in coring, the most important one of which is the depth it can be safely used without twisting or fatigue (Fig. 13.5).

13.5   Underbalanced Coring

The term *underbalance* in drilling is defined as the condition, wherein the bottom-hole pressure is less than the formation pore pressure. Normally, underbalanced coring is only recommended for very low-pressured formations, which have a high potential for formation damage or in case of preparing to core the rock in a loss zone (Keith et al. 2016). A carefully designed underbalanced system aids to mitigate the core damage (due to mud filtrate invasion, clay swelling, phase trapping, etc.) in addition to an increase in the coring/core-drilling rate (i.e., less coring time). Typically, an underbalanced coring system using the outer tube OD of 5¾-in. can be utilized to retrieve a 3½-in. core in the 6 1/8-in. hole size. The coring fluid can be air, gasified liquids, foam, mist, etc. In case of using air, the blooie line must be observed for dusting prior to beginning to core the rock, or after each connection. When dusting was not observed (i.e., clean downhole), it is time for coring. Typically, short core barrel length (i.e., one core barrel joint or 30 ft) is recommended.

Application of underbalanced coring has its safety issues and is not possible with other coring systems such as gel coring, sponge coring, full-closure coring, etc.

References

Deschamps, B., S. Desmette, R. Delwiche, R. Birch, J. Azhar, M. Naegel., and P. Essel. 2008. *Drilling to the Extreme: The Micro-Coring Bit Concept*, SPE 17920-MS OTC. Presented in IADC/SPE Asia Pacific Drilling Technology Conference and Exhibition, Jakarta, Indonesia, August, 25–27.

Fleckenstein, W.W., and A.W. Eustes. 2003 *Novel Wireline Coring System*, SPE 84358 MS. Presented at the SPE Annual Technical Conference and Exhibition, Denver, Colorado, October 5–8.

Keith, C.I., A. Safari, K.L. Aik, M. Thanasekaran, and M. Farouk. 2016. *Coring Parameter optimization-The Secret to Long Cores*. Presented at the OTC Conference, Kuala Lumpur, Malaysia, March 22–25.

Leising, L.J., and E.A. Rike. 1994. *Coiled-Tubing Case Histories*. Presented at the IADC/SPE Drilling Conference, Dallas, Texas, February 15–18.

Miller, J.E., and D.P. Huey. 1992. *Development of a Mud-Motor-Powered Coring Tool*, OTC 6865. Presented at the 24th Annual OTC in Houston, Texas, May 4–7.

Rathmell, J.J., B.S. Wilton, B.A. Gale, D.A. Bell, G.A. Tibbitts, and D.J. Bobrosky. 1998. *Development and Application of PDC Core Bits for Downhole Motor Low-Invasion Coring in the Arab Carbonates*, SPE 36263-PA. Published in the SPE Drilling and Completion Journal, March, pp. 56–65.

Shinmoto, Y., S. Kodama, B. Deschamps. 2012. *Evaluation of the Micro-Core Quality Using Drilling Mechanics Data*, SCA2012-38. Presented at the International Symposium of the Society of Core Analysts, Aberdeen, Scotland, UK, August 27–30.

Storms, M.A., S.P. Howard, D.H. Reudelhuber, G.L. Holloway, P.D. Rabinowitz, and B.W. Harding. 1991. *A Slimhole Coring System for Deep Oceans*, SPE 21907-MS. Presented at the SPE/IADC Drilling Conference, Amsterdam, the Netherlands, March 11–14.

Chapter 14
Core Handling

14.1 Introduction

When the core barrel containing the core column reaches the surface, the inner tube should be safely handled so that the contained core undergoes the least damage (inferred from Skopec 1992, 1994; Hettema et al. 2002; Shafer 2013; Owens and Evans 2013). To achieve this, in the beginning, a systematic and careful planning is a requirement, which also involves selection of the proper tools in the core barrel (such as disposable inner tubes, NRITS) and the surface handling tools. Next, the

© Springer International Publishing AG, part of Springer Nature 2018
R. Ashena and G. Thonhauser, *Coring Methods and Systems*,
https://doi.org/10.1007/978-3-319-77733-7_14

proper communication between all the involved, already-trained personnel is required to guarantee the success of the operation.

Therefore, in this chapter, following the discussion on the required handling tools, the procedure is described step-by-step. Finally, some recommendations are presented to prevent the handling-related core damage.

14.2 Handling Tools

In general, core handling tools are utilized for assembling, dissembling, and laying down of the coring tools, including the core barrel/inner tube. Typically, the objective of coring, the type, and size of the core barrel are the factors that must be considered for the selection of the handling tools. Some of these tools consist of the drop-ball pick-up tool (Fig. 14.1), the inner tube lift bail/sub (Fig. 14.2), the bit breaker (Fig. 14.3), lay-down tools (Fig. 14.4), and splitters (Fig. 14.5).

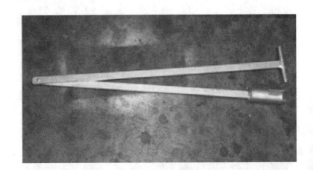

Fig. 14.1 *Drop-ball pick-up* tool to extract the ball stuck inside the inner tube (published courtesy of Baker Hughes GE)

Fig. 14.2 The inner tube *lift bail*, which is used for lifting the inner tube out of the hole (published courtesy of Baker Hughes GE)

Fig. 14.3 The bit breaker,
used to remove the core bit
(published courtesy of Baker
Hughes GE)

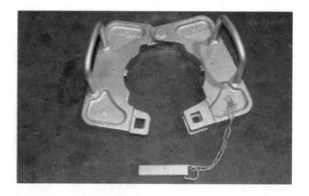

14.3 Procedure

Generally, core handling is composed of three phases:

- Lifting, which includes receiving the inner tube or the core barrel at the rig floor.
- Moving and laying down the inner tube joints from the rig floor to the pipe-deck on the ground.
- Surface gamma-ray logging and cutting the inner tube joints into sections.
- (If necessary) conducting a quick core analysis at the rig site.
- After possible core preservation, loading the cores in boxes, and transportation to the city core laboratory.

A standard handling procedure is essentially required to be followed, depending on the coring method (wireline continuous method does not include some of the below items):

1. In case of conventional coring, the outer tube joints must be first unscrewed and removed. [*not for the wireline method*]
2. Break and remove all crossover subs on top of the core barrel (above the inner tube and between the outer and inner tubes). [*not for the wireline method*]
3. Remove the debris from the top of the core barrel.
4. Extract or retrieve the drop-ball from the top of the inner tube by inserting the *drop-ball pickup/retrieving tool*. Check if the ball has been eroded.
5. Connect/make-up the *lift bail/sub* (hung on the tugger line) to the inner tube (Fig. 14.6).
6. Unscrew and remove the core bit. [*not for the wireline method*]
7. Breakout the *top sub/safety joint* from the outer tube (depending on the core barrel size[1] and type used, between 6 and 13 turns are required to unscrew the top sub). [*not for the wireline method*]

---

[1]Examples of core barrel sizes are: 3 ½ * 1 ¾, 4 1/8 * 2 1/8, 4 ¾ * 2 5/8, 5 ¾ * 3 1/3, 6 ¼ * 4, 6 ¾ * 4, 7 5/8 * 5 ¼. It is noted that the first number is the OD of the outer tube and the second is the ID of the inner tube.

Fig. 14.4 Some standard core lay-down tools and methods (published courtesy of Baker Hughes GE): **a** *Four-point (spreader) beam-sling*, which uses four ropes to attach, lift, and transfer the inner tube from lay-down-frame (LDF) to the pipe deck. **b** The inner tube is fastened to the clamp from the bottom and attached to a cradle using two ropes near the top and bottom of the barrel. **c** A handling cradle (called lay-down frame, LDF) is used to transport each inner tube from the vertical position on the rig floor to the horizontal position on the pipe decks/ground. The inner tube is clamped to LDF using some lines, cables, or latches. It prevents the cores from bending or undergoing bending stresses during transportation

(a)

(b)

(c)

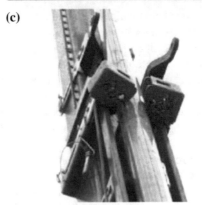

(a)

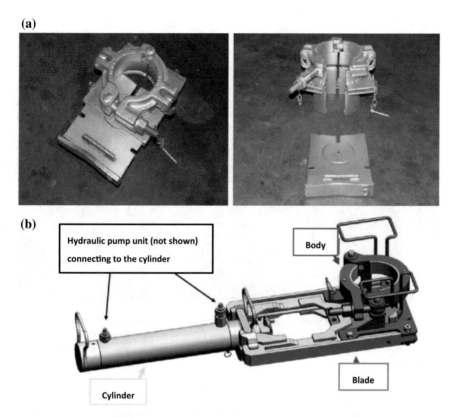

(b)

Hydraulic pump unit (not shown) connecting to the cylinder

Body

Cylinder

Blade

Fig. 14.5 a The mechanical core splitter/guillotine with its two main components, one for clamping the inner tube, and the other one for cutting the inner tube from the bottom, **b** the hydraulic core splitter (published courtesy of Baker Hughes GE)

8. In lieu of the top sub, thread a sub on top of the outer tubes and tighten with tongs. [*not for the wireline method*]
9. Using the rig elevators, lift the outer tube until the inner tube is exposed. [*not for the wireline method*]
10. As the inner tube has become exposed, install the *inner tube clamp/dog collar* or *guillotine clamp* close to the pin connection (of the top joint).
11. Place the *guillotine/splitter knife or blade* at the bottom of the top inner tube, twist the steel inner tube or backoff the NRITS for opening a window in order to cut.

   – It is not possible to cut steel inner tubes (which are non-disposable). Cutting the inner tubes using the splitter means cutting only the disposable inner liners (if used, such as the aluminum or fiberglass liners) and the contained core. Traditionally, to expose the disposable inner tube before cutting, the joints of the steel inner tubes were twisted (by unscrewing the pin and box connection between the 30-foot joints), which could mechanically damage

(a)   **(b)**

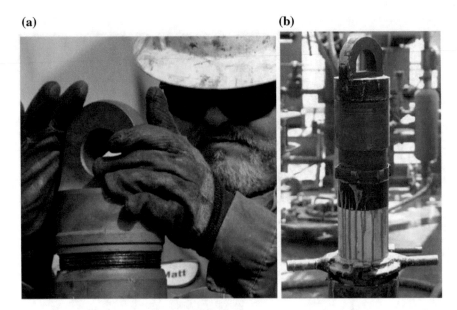

Fig. 14.6 In conventional coring, the lift bail/sub is connected to the top of the inner tube and is raised using the tugger line to pull the next inner tube out of the outer tube. Now, the same cutting and lay-down process can be repeated for this inner tube joint (published courtesy of Baker Hughes GE)

Fig. 14.7 NRITS is used to connect two inner tube joints (each 30 ft/9.5 m), which can be opened by latching the splitter *skirt* on the NRITS collar and then partial backingoff of the torque lock (as part of the NRITS) to open a window for cutting by the splitter cutting-knife. This, first, removes the necessity of twisting of the normal pin and box connections which could dramatically damage the core (torque-induced fractures, etc.) and second, it provides the window for cutting the disposable inner liner/core by the splitter (published courtesy of Baker Hughes GE)

the sample (Hettema et al. 2002). Recently, a connection named *Non-Rotating Inner Tube Stabilizer NRITS* (shown in Fig. 14.7) is commonly used as part of the inner tube, which connects the inner tube joints/sections (particularly for extended or long coring runs, e.g., 270 ft).

Fig. 14.8 Mechanical core splitter/guillotine, which is mechanically turned between the inner tube joints in order to cut the disposable inner liner/core (published courtesy of Baker Hughes GE)

Using NRITS, just by a partial *back-off* of the *NRITS collar*, two halves of the NRITS are separated (i.e., a *window* is exposed) so that the guillotine blade/splitter knife can easily cut the core.

– Splitters/guillotines can be mechanical (Fig. 14.8) or hydraulic (Figs. 14.9 and 14.10).

12. When the top inner tube joint has been cut, it is secured by a *core cup* and *cushion* and the core splitter/guillotine is detached from the inner tube (Fig. 14.11a).

13. Each inner tube joint containing the core is placed inside an *LDF* (Fig. 14.12) and raised on the air tugger *line* or even the crane (e.g., with two lines fastened to the top and bottom of the inner tube). Each inner tube joint containing the core is transferred from the vertical position on the rig floor to the horizontal position on the special pipe-deck (on the ground) (Fig. 14.13).

– To transport the inner tube from the rig floor to the ground level, the LDF is laid-down on the catwalk.

– Most LDFs are equipped with rollers for easier transport of the inner tubes.

14. Still, the next inner tube joints must be cut. Use the tugger line again to lift the *lift bail/sub* about 30–40 ft/9.5–12 m up into the derrick in order to raise the next joint of the inner tube above the rotary table. Then, this joint will be similarly cut and laid-down from the rig floor to the ground.

    Note: After all the inner tube joints were laid-down on the surface pipe-decks if further coring is required the outer tubes are reloaded with new inner tubes to be run downhole.

Fig. 14.9 A hydraulic core splitter: **a** the splitter is around the inner tube (its clamps are in latched state), **b** the hydraulic pump is installed to the splitter, and **c** cutting the disposable inner liners between the joints (published courtesy of Baker Hughes GE)

(a)

(b)

(c)

Fig. 14.10 The inner tube joint (with NRITS between every two joints) has been cut by the splitter (published courtesy of Baker Hughes GE)

(a)   **(b)**

Fig. 14.11 a After cutting the core at the bottom of the inner tube joint, it is secured by placing the cap and cushion on the bottom (of the upper half of NRITS), and the splitter is detached or removed from the inner tube joint, **b** the inner tubes to be cut, the left is a cut and secured inner tube, which is going to be loaded and attached to the LDF for transportation to the pipe-decks/ ground (published courtesy of Baker Hughes GE)

(a)   (b)

Fig. 14.12 a Lay-down frame/cradle (LDF) reaches the rig floor, and **b** is positioned just next to the cut inner tube section to stabilize and transport the cut inner tube (containing the core) from the rig floor to the pipe-deck on the ground (published courtesy of Baker Hughes GE)

Fig. 14.13 Transferring the LDF from the rig floor to the pipe-deck on the ground (published courtesy of Baker Hughes GE)

Fig. 14.14 Gamma-ray logging of the core at the surface, which is done before cutting the inner tube joints into one-meter sections (published courtesy of Baker Hughes GE)

15. Obtain a *surface gamma-ray log*[2] from the cores (Fig. 14.14) prior to cutting each core joint (30 ft/9.5 m) into one-meter sections. This is done to:

    - Evaluate the lithology and identify the interval of the zone of interest (reservoir formation, shale intervals, etc.), and deciding if further coring is required.
    - Depth-match or correlate the core depth with the full-set wireline data (which may be probably run after coring).
    - Identify intervals with higher importance for sampling and core analysis.

    Note: In case of using *LWC system* (Logging-While-Coring), surface gamma-ray logging and the wireline logging (which may be run after coring) can be ignored.

16. Immediately after the surface logging, using the *air saw*, cut the inner tube joints containing the core samples (Fig. 14.15) into one-meter sections and mark them.

    - The air saw could be powered mechanically or by an external generator.
    - Mark the core sections from top to indicate their tops and the length. Place rubber caps on both ends of each cut core section.

---

[2]The surface gamma logging service was developed after the 1980s and is still popular. It is used prior to stabilizing and preserving the core in order to obtain reservoir data particularly for quick identification of pay zones.

(a)   **(b)**   **(c)**

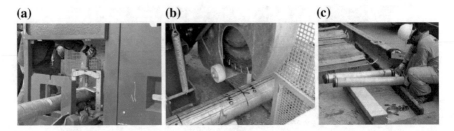

Fig. 14.15 a The inner tube joints are placed horizontally by the LDF next to the saw (just before cutting), **b** cutting the core using the saw, **c** the ends of the inner tube sections are fastened with the *end cap clips* (published courtesy of Baker Hughes GE)

- For cutting fiberglass inner tubes, use the diamond impregnated saw while using a fluid coolant for its cooling. For cutting aluminum inner tubes, use the tungsten carbide saw.

  Note: In modern coring, the inner tube is usually out of steel (not aluminum or fiberglass) and a liner (non-spilt or split) may be used inside the inner tube to contain the core.

17. Clean the inner tube joints and then mark them with cut-lines.

    - Write the top and bottom depths for each one meter.
    - The lines perpendicular to the pipe axis show the location of the cutting sections.

18. For more efficiency and decision-making purposes, sometimes RCAL and SCAL are required quickly for rapid decision-making (depending on the job plan). In such cases, a quick wellsite core analysis can be performed (depending on the coring objectives):

    - This requires a core analysis unit, but it is a proper option particularly in remote areas or when decision-making depends on core analysis.
    - The typical analyses that can be performed include microscopic examination of the lithology, core plugging, e.g., with plugs of 1.5-in. diameter (Nnoaham and Marchall 2010), identification of the zone of interest using the surface gamma-ray and wellsite core analysis, fluorescence determination with ultra-violet light, determination of porosity and permeability, etc.

19. Place the inner tube sections (each one-meter) into boxes for backload and shipment (Fig. 14.16):

    - The box can be wooden, rubber, or metal.
    - *Core cubes* can be also used for better protection of the cores.

(a)   **(b)**

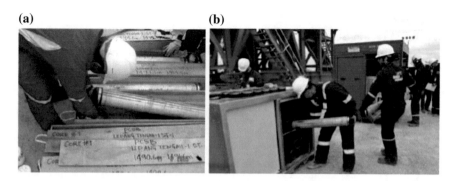

Fig. 14.16 Placing the inner tube sections in **a** wooden boxes, or **b** core cubes for backload and transportation to the city core analysis lab (published courtesy of Baker Hughes GE)

> **Note**: In case there is a need for preservation of the core(s), first preserve them before placing them into boxes. Avoid washing the cores with water. Discussion on the preservation methods is out of the scope of this book.

20. Transport the core(s) to the city core analysis laboratory.

   – Avoid unsafe transportation means of the core (Fig. 14.17).
   – Rubber core boxes/containers can keep the core in better condition than the wooden ones. An electronic shock monitor can be mounted in each box in order to record or monitor any possible damage to the cores.

Fig. 14.17 Inappropriate transportation of the inner tubes to the central laboratory. In this example, it is seen that even cutting of the long inner tube into sections has not been done (Hettema et al. 2002)

Attention
It is noted that in case of pressure coring, a special core barrel handling is required (Al Neaimi et al. 2014). Therefore, the above steps 10–12 are replaced by steps 9–14 in Sect. 11.5, then, it is continued from step number 13 to the end.

14.4   Recommendations

It is inferred in this chapter that the handling-induced core damage generally occur due to:

1. Rotating the inner tube (normal pin-box connection) during unscrewing the connections.
2. Cutting the core by splitters or saws.
3. Laying-down the inner tube joints containing the core from the vertical position (the rig floor) to the horizontal (the pipe-decks).

Systematic and careful planning and execution of handling practices including good communication among all the persons involved can guarantee prevention of damage to the core during its transportation. Therefore, considering the discussions of the chapter, there are several handling-related recommendations to mitigate core damage:

- Use non-rotating inner tube stabilizers (NRITS) between the inner tube joints to minimize the rotation of the steel inner tube joints (while opening their threads), which can cause severe damage at the threaded connection points.
  Note: In case NRITS have not been used (for many reasons), cut the inner tubes a few centimeters above the threaded connections to eliminate the core damage (Hettema et al. 2002).
- Use hydraulically-powered core splitters/guillotine, instead of mechanical ones, for cutting the core. This prevents damage to the core sample.
- Use a disposable inner liner (preferentially aluminum[3]) inside the steel inner tube to highly maintain the core quality during handling, particularly in unconsolidated formations (Whitebay 1986; API 1998; Hettema et al. 2002; Nnoaham and Marchall 2010; Owens and Evans 2013). Using easily-openable liners such as *half-moon or 2/3-moon aluminum liners or laser/plasma-cut liners* is strongly recommended particularly in unconsolidated formations.
- Use modern H-beam lay-down frames (LDF with latches, instead of ropes) for safer lay-down of the core with less bending (API 1998; Hettema et al. 2002; Owens and Evans 2013).

---

[3]Liners can be also fiberglass or PVC, but aluminum liners are less prone to flex than their equivalents (Hettema et al. 2002).

- As soon as the inner tubes containing the cores reach the pipe-decks on the ground, quickly start cleaning, marking, and surface gamma-ray spectroscopy, before cutting the inner tube joints into one-meter sections.
- Using air-powered saw is recommended for cutting the inner tubes into sections while the core is kept in the cradle and is continuously run into the saw location by rollers.
- Avoid washing the core with water, just preserve it before loading.
- Core cubes are recommended for loading rather than wooden boxes, before shipment to the central core analysis lab.

References

Al Neaimi, M.A., A.S. Tee, D. Boyd, R. Al Shehhi, E.A. Mohamed, K.M.N Namboodiri, H.A. Junaibi, M. Al Zaabi, H. Al Braik, S. Ftes, A.K. Medjiah, A. Farouz, B. Gao, M. Gay, S.M. Tariq, B. Fudge, P. Collett, T. Gill, A. Abdul-hamid, B. Schipper, and M. Sheldon. 2014. *Acquisition of an Elevated Pressure Core in a Gas Flooded Carbonate Oil Reservoir: Design and Operational Challenges*, SPE 171815 MS. Presented at the Abu Dhabi International Petroleum Exhibition and Conference, November 10–13.

American Petroleum Institute (API). 1998. *Wellsite Core Handling Procedures and Preservation*. Recommended Practice-40 (RP-40).

Hettema, M.H.H., T.H. Hanssen, and B.L. Jones. 2002. *Minimizing Coring-Induced Damage in Consolidated Rock*, SPE-78156-MS. Presented at the SPE/ISRM Rock Mechanics Conference, 20–23 October, Irving, Texas

Nnoaham, I., and D. Marchall. 2010. *Improving Deepwater core Recovery and Onsite handling Through Strategic Coring Protocol*, SPE 140628 MS. Presented at the 34th Annual SPE International Conference and Exhibition, Tinapa-Calabar, Nigeria.

Owens, J., and Evans, R.D. 2013. *Extreme HPHT Coring, Handling and Analysis: Securing the Value Chain*, Presented at the International Symposium of the Societ of Core Analysts, Sep. 16–19, Napa Valley, California, USA.

Shafer, J. 2013. Recent Advances in Core Analysis. *Petrophysics* 54(6): 554–579.

Skopec, R.A., M.M. Mann, D. Jeffers, and S.P. Grier. 1992. Horizontal Core Acquisition and Orientation for Formation Evaluation. *SPE Drilling Engineering Journal*, March, pp. 47–54.

Skopec, R.A. 1994. Proper Coring and Wellsite Core Handling Procedures: The First Step Toward Reliable Core Analysis, SPE 28153-PA. *Journal of Petroleum Technology (JPT)*.

Whitebay, L.E. 1986. *Improved Coring and Core-Handling Procedures for the Unconsolidated Sands of the Green Canyon Area, Gulf of Mexico*, SPE 15385 MS. Presented at the SPE Annual Technical Conference and Exhibition, 5–8 October, New Orleans, Louisiana.

Chapter 15
Coring Providers and Patents

15.1 Introduction

In the first part of this chapter, detailed specifications and information regarding the dimensions of the core samples, hole size, and working parameters corresponding to some important coring tools, which are provided by the coring providers (both in the petroleum and mining industries) are presented. In the second part of the chapter, some important patent information regarding the current coring systems has been listed.

This chapter contributes to obtaining an understanding of the dimensions of the available coring tools in the market, and also the pioneers and patents which aided the development of the coring methods and tools.

15.2 Coring Providers and Specifications

The coring providers are subdivided into the petroleum and mining companies.

© Springer International Publishing AG, part of Springer Nature 2018   213
R. Ashena and G. Thonhauser, *Coring Methods and Systems*,
https://doi.org/10.1007/978-3-319-77733-7_15

15.2.1   Petroleum Sector

In this section, the specifications and detailed information for some important coring tools provided by some main petroleum providers have been listed in Tables 15.1, 15.2, 15.3, 15.4, 15.5, 15.6, 15.7, 15.8, 15.9, 15.10, 15.11, 15.12, 15.13, 15.14, 15.15, and 15.16.

The provider companies consist of *Baker Hughes GE, NOV, Reservoir Group (formerly ALS Corpro)*, and *Halliburton*. Some coring systems of *Baker Hughes GE* are: conventional coring, wireline continuous coring (*CoreDrill*), low-invasion system (*CoreGard*), gel coring (*CoreGel*), sponge coring (*SOr*), antijamming (*JamBuster*), and full-closure (*HydroLift*). Some coring tools of NOV are: conventional coring, wireline continuous coring, low-invasion coring, sponge coring (*Enhanced Oil Saturation, EOS*), antijamming (*JamTeQ*), and full-closure (*DuraClose*). Some of *Reservoir Group*'s coring systems are: conventional coring, wireline continuous coring, low-invasion system, sponge coring, antijamming, and full-closure. Some of the Halliburton's coring systems are: conventional coring (RockStrong), wireline continuous coring (*Lathchless*), full-closure coring, and side-wall pressure coring (*CoreVault*). In terms of commerciality of the tools, it is noted that usually the premium coring products are only rented by provider companies, however, the older tools can be sold.

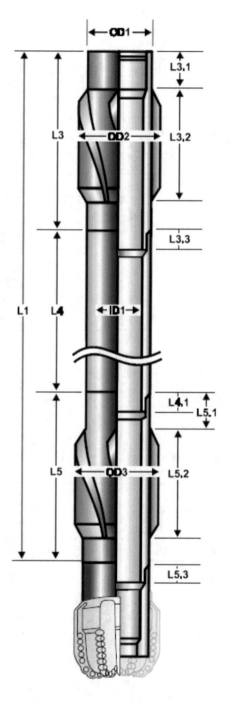

(published courtesy of Baker Hughes GE)

Table 15.1 Detailed specifications of conventional coring (published courtesy of Baker Hughes GE)

Baker Hughes GE outer barrel system	250P series outer-barrels		HT-series outer barrels					
Core barrel type	5.50 in. 250P	6.25 in. 250P	HT10	HT12	HT30	HT30 Max	HT40	HT60
Conveyance	Drillpipe	Drillpipe	Drillpipe/coil tubing	Drillpipe/coil tubing	Drillpipe/coil tubing	Drillpipe	Drillpipe	Drillpipe
Barrel diameter	5.50 in. (139.7 mm)	6.25 in. (158.8 mm)	4.75 in. (120.6 mm)	5.125 in. (130.2 mm)	6.75 in. (171.4 mm)	7.25 in. (184.1 mm)	8.00 in. (203.2 mm)	9.50 in. (241.3 mm)
Top connection	NC38	4.50 in. (114.3 mm) API FH	NC38	NC38	NC50	NC50	6.625" (168.3 mm) API Reg	6.625" (168.3 mm) API Reg
Hole size	6.125 in. (155.6 mm) and upwards	6.75 in. (171.4 mm) and upwards	5.75 in. (146.0 mm) and upwards	6.00 in. (152.4 mm) and upwards	8.00 in. (203.2 mm) and upwards	8.875 in. (225.4 mm) and upwards	9.00 in. (228.6 mm) and upwards	12.25 in. (311.1 mm) and upwards
Conventional core size	3.50 in. (88.9 mm)	4.00 in. (101.6 mm)	2.625 in. (66.7 mm)	3.00 in. (76.2 mm)	4.00 in. (101.6 mm)	4.50 in. (114.3 mm)	4.75 in. (120.6 mm)	5.25 in. (133.3 mm)
Core barrel length	Multiples of 30 ft (9.14 m)	Multiples of 30 ft (9.14 m)	Multiples of 30 ft (9.14 m)	Multiples of 30 ft (9.14 m)	Multiples of 30 ft (9.14 m)	Multiples of 30 ft (9.14 m)	Multiples of 30 ft (9.14 m)	Multiples of 30 ft (9.14 m)
Make-up torque	7000 lbf/ft (9500 N m)	9000 lbf/ft (12,750 N m)	10,000 lbf/ft (13,558 N m)	12,000 lbf/ft (16,269 N m)	30,000 lbf/ft (40,675 N m)	30,000 lbf/ft (40,675 N m)	40,000 lbf/ft (54,233 N m)	60,000 lbf/ft (81,349 N m)
Yield torque (kft lbs)	11,100 lbf/ft (15,050 N m)	11,800 lbf/ft (16,000 N m)	22,100 lbf/ft (29,960 N m)	19,700 lbf/ft (26,710 N m)	55,300 lbf/ft (74,975 N m)	46,800 lbf/ft (63,450 N m)	64,900 lbf/ft (87,990 N m)	111,800 lbf/ft (151,580 N m)
Yeild overpull	300,000 lbs. (1330 kN)	350,000 lbs. (1560 kN)	292,000 lbs. (1299 kN)	328,500 lbs. (1461 kN)	328,500 lbs. (1461 kN)	591,000 lbs. (2630 kN)	663,000 lbs. (2950 kN)	1088,000 lbs. (4840 kN)
Max. DLS (°/30 m)	3.9°	6.0°	6.3°	5.1°	5.1°	5°	4.2°	3.9°

(continued)

Table 15.1 (continued)

Baker Hughes GE outer barrel system	250P series outer-barrels		HT-series outer barrels					
Max. DLS (°/30 m)	7.9°	12°	14.1°	12.6°	12.6°	14°	12°	7.8°
Jambuster system	No	Yes	Yes	Yes	Yes	Yes	Yes	Yes
Lasercut system	No	No	No	Yes	Yes	Yes	No	No
Hydrolift system	No	No	No	No	Yes	No	No	Yes
Sponge SOr system	No	No	No	No	Yes	No	No	No
Flapper catcher	No	Yes	Yes	No	Yes	Yes	No	Yes
NRITS available	No	No	Yes	Yes	Yes	Yes	Yes	Yes
Oriented coring	Yes	Yes	No	Yes	Yes	Yes	Yes	Yes
HPHT capability	No	No	Yes	Yes	Yes	Yes	Yes	Yes

The data below are generic information on the tools provided earlier for these services

Table 15.2 Detailed specifications of conventional coring (published courtesy of NOV)

Barrel type	Durateq[a]	P2500[b]	P2500	P2500	P2500	Durateq	Durateq
Barrel size	4¾ * 2 5/8	4¾ * 2 5/8	5 * 3	6¼ * 4	6¾ * 4	7 * 4	9½ * 5¼
Hole size (in.)	5½–6¾	5 ½–6¾	5 7/8–6¾	6¾–7 7/8	8 ½–8 3/4	7 7/8–9 7/8	12¼
Core size (in.)	2 5/8	2 5/8	3	4	4	4	5¼
Core barrel L (m)	Multiples of 9.5	9, 18, 27	9, 18	9, 18, 27	8, 18, 27	Multiples of 9.5	Multiples of 9.5
Max core barrel L (ft)	800	90	90	90	90	800	800
Max no. Joints	27	3	3	3	3	27	27
Outer tube OD (in.)	4¾	4¾	5	6¼	6¾	7	9½
Outer tube ID (in.)	3¾	3¾	4	5 1/8	5 3/8	5½	7
Outer tube wall δ (in.)	½	½	½	0.56	0.68	¾	1¼
Hole-outer tube δ (in.)	0.37–1	0.37–1	0.43–0.87	¼–0.81	0.87–1	0.43–1.43	1.375
Inner tube OD (in.)	3 3/8	3 3/8	4¾	4¾	4¾	4¾	6
Inner tube ID (in.)	2 7/8	2 7/8	4¼	4¼	4¼	4¼	5½
Inner tube wall δ (in.)	¼	¼	¼	¼	¼	¼	¼
Steel ball size (in.)	1¼	1¼	1¼	1¼	1¼	1¼	1¼
Pressure relief plug (in.)	1	1	1	1	1	1	1
OD of DP[c] (in.)	4½–5	4½–5	4–4½	4½–5	4½–5	4½–5	4½–5
Drift diameter of DP	3.5–4.3	3.5–4.3	3.2–3.8	3.5–4.3	3.5–4.3	3.5–4.3	3.5–4.3
Top connection type[d]	NC	IF	IF	FH	IF	NC	Reg
Safety joint/top sub ID (in.)	~2.75	~2.75	~2.75	~2.75	~3.25	~3.25	3.25
O.T. Make-up torque (K ft lb)	10	4.7	8.6	8.6	9.4	31.5	62

(continued)

Table 15.2 (continued)

Barrel type		Durateq[a]	P2500[b]	P2500	P2500	P2500	Durateq	Durateq
O.T. Yield torque (K ft lb)		15.5	5.8	12.5	11.8	13.9	48.5	103.25
O.T. Max pull (K lb)		470	137	253	350	276	1.037	1.051
Max Al I.T.[e] Make-up torque (K ft lb)		0.6	0.6	~0.6	~1	1	~1	2
I.T. Yield torque (K lb)		1	1	1	2.1	2.1	2.1	4.2
I.T. max pull (K lb)		35	35	35	50	50	50	105
Operational parameters	WOB (K lb)	~5–15	4–13	4–16	4–16	6–15	~15–45	~20–90
	RPM	30-motor RPM	30–100	30–100	30–100	30–100	30-motor RPM	30-motor RPM
	GPM	80–150	175–225	175–250	300–525	250–350	90–230	110–260
Possibility of full-closure coring		Yes	–	–	–	–	Yes	Yes
Possibility of directional/motor coring		Yes	–	–	–	–	Yes	Yes

[a]They are equipped with high strength and double shouldered threads (durable threads)

[b]Older version of high torque barrel

[c]Conventional drill pipes are used (1) with OD of 5-in., 19.5 lb/ft, (2) OD: 4½-in., 16.6 lb/ft, (3) OD: 3½-in., 9.5 lb/ft, and (4) OD: 2 7/8-in., 6.85 lb/ft

[d]The size corresponds to the ID of box connection of O.T.

[e]Aluminium inner tube with maximum strength

Table 15.3 Specifications of wireline continuous coring systems (courtesy of NOV)

	NOV	
	Velocity-525	Corion
Barrel size	5¼ * 3½	6½ * 3/3½
Hole size (in.)	5 7/8–7 7/8	7 7/8–8 3/4
Core size (in.)	3½	3 or 3½
Max depth (m)	4200	5400[a]
Max core barrel L (m)	9–55 m	7.4, 9, 18, 27 m
Max core barrel L (ft)	180	90
Max no. joints	6	3
Outer tube OD (in.)	5¼	6½
Outer tube ID (in.)	4 3/8	4¼
Outer tube wall δ (in.)	0.43	1.125
Hole-outer tube δ (in.)	0.31–1.31	0.68–1.12
Inner tube OD (in.)	~4 1/8	3.8
Inner tube ID (in.)	~3 7/8	3.6–3.5
Inner tube wall δ (in.)	0.18	0.1–0.15
Steel ball size (in.)	1¼	1¼
Bit insert/plug D (in.)[b]	3, 3½	3½
Pressure relief plug (in.)	1	1
OD of DP (in.)	5–5½	5–5½
Drift diameter (of DP) (in.)	4¼[c]	4¼
Possibility of jars	Yes	Yes
O.T. make-up torque (K ft lb)	8.736	26
Yield Torque (K ft lb)	13.44	44.8
Max pull/tensile yield (K lb)	28.185	926

(continued)

Table 15.3 (continued)

Operational parameters		NOV	Corion
		Velocity-525	
Operational parameters	WOB (K lb)	4–13	4–24
	RPM	30–100	30–150
	GPM	17–250	175–300
Max DLS (°/100 ft)	Rotating	3.6	3.6
	Non-rotating	7.2	7.2
Max inclination angle (°)		30–50	50
Wireline trip speed (m/min)		60–120	60–120
Inner tube assembly weight, coring mode/inner drilling assembly weight, drilling mode (lb) for 30 ft length		~180/750	~180/750
Top connection type		9VS[d]	Propr.5DDD
Application of vented inner tube		No	No
Possibility of oriented coring		Yes	Yes
Possibility of logging-while-coring		No	No
Possibility of low-invasion coring		Yes	Yes
Possibility of gel coring		No	No
Possibility of full-closure coring		No	No
Possibility of antijamming		No	No

[a]World records in Poland
[b]Equal to the core diameter
[c]DP with 5-in. OD and nominal weight of 16.26 lb/ft or 5½-in. DP (refer to DP specifications, RP-7G: drill stem design and operating limits)
[d]Vam standard

Table 15.4 Conventional and soft-pro coring systems (published courtesy of Reservoir Group)

Barrel size		4¾ * 2 9/16	5 1/8 * 3	5½ * 3½	7 1/8 * 4	8 5/8 * 5¼	9½ * 5¼
Hole size (in.)		5¾–6½	5 7/8–7	6–7	8–9¼	9¼–11	11–12¼
Core size (in.)		2 9/16	3	3½	4	5¼	5¼
Type		Conv./Softpro	Conv.	Conv.	Conv./Softpro	Conv./Softpro	Conv./Softpro
Core barrel L (m)		Multiples of 9	Multiples of 9	Multiples of 9	Multiples of 9	Multiples of 9	Multiples of 9
Outer tube OD (in.)		4¾	5 1/8	5½	7 1/8	8 5/8	9½
Outer tube ID (in.)		3¾	4 1/8	4½	5 5/8	7	7
Outer tube wall δ (in.)		½	½	½	¾	0.812	1¼
Hole-outer tube δ (in.)		½–1¾	¾–0.937	¼–¾	0.43–1.06	0.31–1.18	¾–1.37
Inner tube OD (in.)		–	–	–	–	–	–
Inner tube wall δ (in.)		–	–	–	–	–	–
Steel ball size (in.)		1¼	1¼	1½	1½	1 7/8	1 7/8
OD of DP (in.)		3½–4½	4½–5	4–4½	5½[a]	5½	5½
ID of DP (in.)		–	–	–	4.778	4.778	4.778
Drift diameter of DP		3.5–4.3	3.5–4.3	3.2–3.8	3.5–4.3	3.5–4.3	3.5–4.3
Core barrel thread type		Corpro HD	Corpro HD	Corpro HD	Corpro HD	Corpro HD	Corpro HD
Top connection type		3½ IF	3½ IF (NC38)	3½ IF (NC38)	4½ IF	6 5/8 REG	6 5/8 REG
O.T. Make-up torque (K ft Ib)		7.5	12	13	30	35	45
O.T. Max torque (K ft Ib)		25	20	22	70	75	120
O.T. Max pull (K Ib)		300	350	380	860	960	1550
Operational parameters	WOB (K Ib)	~5–15	4–13	4–16	4–16	6–15	~20–90
	RPM	30-motor RPM	30–100	30–100	30–100	30–100	30-motor RPM
	GPM	80–150	175–225	175–250	300–525	250–350	110–260
Possibility of full-closure coring		Yes	–	Yes	Yes	Yes	Yes
Possibility of directional/motor coring		Yes	Yes	Yes	Yes	Yes	Yes
Possibility of oriented coring		Yes	Yes	Yes	Yes	Yes	Yes

[a]OD: 5½-in., 21.9 Ib/ft, ID: 4.778, tool joint OD: 6½-in., tool joint ID: 4½-in.

Table 15.5 Specifications of wireline continuous coring with pressure coring[a] (published courtesy of Reservoir Group)

Barrel size	4¼ * 3	8 3/8 * 4
Hole size (in.) with wireline	7 7/8	–
Hole size (in.) with conventional drill pipes	6¼	8 3/8
Core size (in.)	3	4
Standard core barrel L (m)	9	6
Max core L (m), for each run	3	3.6
Max total core L (m)	36	36
Canister L (m) (pseudo-pressure coring)	4.5–9	4.5–9
Outer tube OD (in.)	4 ¼	5 ¼
Outer tube ID (in.)	–	–
Outer tube wall δ (in.)	–	–
Hole-outer tube δ (in.)	1.81	1.56
Inner tube OD (in.)	–	–
Inner tube ID (in.)	–	–
Inner tube wall δ (in.)	–	–
Steel ball size (in.)	1 ¼	1 ¼
OD of DP (in.)	5 ½	5 ½
ID of DP (in.)	4.778	4.778
Inner tube max pressure[b] (psi)	1000	1000
System limiting relief valve pressure (psi)	500	500
Max temperature (°C)	120	120
Wireline working load (K Ib)	6	6
O.T. Typical bit pull (K Ib)	2	4
Possibility of full-closure coring	Yes	Yes
Possibility of oriented coring	Yes	Yes

[a]Called "Quick capture" by Reservoir Group
[b]Also called "*safety relief pressure*" at which the valves in the canister open to vent the gas

Table 15.6 Specifications of sidewall coring systems (published courtesy of Halliburton)

	RSCT	HRSCT[a]
Type	Sidewall, diamond	Sidewall, HTHP (400 °F, 25 K psi)
Hole size (in.)	1.75	2.5
Core size (in.)	Min 5 1/4	Min 5¼
Core L (in.)	0.93	1.5

[a] *Hostile rotary sidewall coring tool*

Table 15.7 General specifications of coring systems (published courtesy of Halliburton)

	Full-closure	Latch less[a]	Corienting
Type	Conventional	Wireline, triple tube	Conventional, oriented
Hole size (in.)	7¾–12¼	5¾–9	5¾–12¼
Core size (in.)	4–5¼	1.7–3	2 5/8, 4, 5¼
Core L (in.)	Multiples of 30 ft	Multiples of 30 ft	Multiples of 30 ft

[a]Proper for unconsolidated reservoir rocks

Table 15.8 Specifications of conventional full-closure coring system [field units] (published courtesy of Halliburton)

Parameter	Full-closure system	
Type	High torque (or heavy duty)	
Barrel size (in.)	6¾ * 4	8 * 5¼
Hole size (in.)	7¾–9	10 5/8–12¼
Core size (in.)	4	5¼
Min core barrel L (m)	9	9
Outer tube OD (in.)	6¾	8
Outer tube ID (in.)	5 3/8	6 5/8
Outer tube wall δ (in.)	0.688	0.688
Hole-outer tube δ (in.)	0.5–1.125	1.938–2.75
Inner tube OD (in.)	4¾	6¼
Inner tube ID (in.)	4¼	5½
Inner tube wall δ (in.)	¼	0.375
1st Steel ball size (in.)	1¼	1¼
2nd steel ball size (in.)	1 5/8	1 5/8
Pressure relief plug (in.)	~1¼	~1¼
O.T. Make-up torque (K ft Ib)	25.8	36.9
Yield torque (K ft Ib)	39	55.5
Max pull/tensile yield (K Ib)	633	783
Max GPM	300	350
Top connection type	4½ IF	6 5/8 Reg

Table 15.9 Specifications of wireline and conventional orienting coring systems (published courtesy of Halliburton)

Parameter	Latch less			Oriented coring		
Type	High torque (or heavy duty)					
Barrel size (in.)	4¾ * 1.71	6¾ * 2	6¾ * 3	4¾ * 2 5/8	6¾ * 4	8 * 5¼
Hole size (in.)	5¾ * 7	7¾ * 9	7¾ * 9	5¾ * 7	7¾ * 9	10 5/8 * 12¼
Core size (in.)	1.71	2	3	2 5/8	4	5¼
Min core barrel L (m)	9	9	9	9	9	9
Outer tube OD (in.)	4¾	6¾	6¾	4¾	6¾	8
Outer tube ID (in.)	3¾	5 3/8	5 3/8	3¾	5 3/8	6 5/8
Outer tube wall (in.)	0.500	0.688	0.688	0.500	0.688	0.688
Inner tube OD (in.)	2.13	2.531	3.75	3 3/8	4¾	6¼
Inner tube ID (in.)	1.810	2.14	3.25	2 7/8	4¼	5½
Inner tube wall (in.)	0.16	0.196	0.25	0.25	0.25	0.375
Bit insert/plug D (in.)	1.71	2	3	–	–	–
Drift diameter (of DP) (in.)	2¼	2 5/8	4 1/8	–	–	–
O.T. Make-up torque (K ft Ib)	9.6	25.8	25.8	9.6	25.8	36.9
Yield torque (K ft Ib)	14.8	39	39	14.8	39	55.5
Max pull (K Ib)	308	633	633	308	633	783
Top Connection Type	3½ IF	4½ IF	4½ IF	3½ IF	4½ IF	6 5/8 Reg

15.2.2   Mining Sector

In this section, the specifications and detailed information for some important coring tools of some main mining providers have been listed. These companies consist of *Sandvik* and *Boart Longyear*.

Thin Kerf Barrels

Note (Sandvik systems)
N/N2/N3: N size is the industry standard. N2 is a *thin kerf*[1] Bit larger OD on bit, which provides a larger core sample and faster cutting speeds. N3 has an extra split tube inside the inner tube to give a better representative core sample,[2] but due to the splits, a smaller OD on the bit yields which gives a smaller core sample. The N and N3 inner tubes are of same size but the N3 system has split tubes inside the inner tube. The N2 inner tube is of larger size. The above also holds for H/H3 and P/P3.

15.3   Coring-Related Patents

In the following Table 15.17, some important patents corresponding to the coring systems are listed.

---

[1]Thin kerf or light weight systems contribute to retrieval of a larger core sample from the same hole size.

[2]In *triple tube system*

Table 15.10 Specifications of standard wireline continuous coring systems [field units] (published courtesy of Sandvik 2013)

System	Hole size	Core D.	Hole-core	Inner tube			ID complete with split tube	Rod (outer tube/barrel) properties					Rod-hole annulus		Depth
				OD	ID	δ		OD	ID	δ	W	V	Volume	δ	DE880
	in.			in.				in.			lb/ft	US gal/ 100 ft	US gal/ 100 ft	in.	ft
A	–	–	–	–	–	–	–	–	–	–	–	–	–	–	–
B (AW casing)	2.36	1.43	0.93	1.69	1.49	0.1	N/A	2.19	1.81	0.2	4.01	13.4	3.2	0.1	13,320
N (BW Casing)	2.98	1.88	1.1	2.19	1.96	0.1	N/A	2.75	2.38	0.2	5.12	23.01	5.4	0.1	10,410
N2	2.98	2	0.98	2.25	2.06	0.1	N/A	2.75	2.38	0.2	5.12	23.01	5.4	0.1	–
N3	2.98	1.78	1.2	2.19		0.2	1.8	2.75	2.38	0.2	5.12	23.01	5.4	0.1	–
H (NW casing)	3.78	2.5	1.28	2.87	2.63	0.1	N/A	3.5	3.06	0.2	7.65	38.27	8.4	0.1	6970
H3	3.78	2.41	1.37	2.87		0.2	2.45	3.5	3.06	0.2	7.65	38.27	8.4	0.1	–
P (HW casing)	4.83	3.35	1.48	3.75	3.49	0.1	N/A	4.5	4	0.3	11.33	65.28	12.5	0.2	4710
P3	4.83	3.27	1.56	3.75		0.2	3.31	4.5	4	0.3	11.33	65.28	12.5	0.2	–

Table 15.11 Specifications of standard wireline coring systems [other units] (published courtesy of Sandvik 2013)

System	Hole size	Core D.	Hole-core	Inner tube				Rod properties					Rod-hole annulus		Depth
				OD	ID	ID split tube	δ	OD	ID	δ	W	Volume	V	δ	DE880[a]
	cm			cm				cm			kg/3 m	l/m	l/m	cm	m
A	–	–	–	–	–	–	–	–	–	–	–	–	–	–	–
B	5.99	3.64	2.35	4.29	3.8	NA	0.25	5.56	4.6	0.48	17.9	1.66	0.4	0.21	4060
N	7.57	4.76	2.81	5.55	4.99	NA	0.29	6.99	6.03	00.48	22.9	2.85	0.67	0.29	3170
N2	7.57	5.07	2.50	5.72	5.24	NA	0.24	6.99	6.03	00.48	22.9	2.85	0.67	0.29	–
N3	7.57	4.51	3.06	5.56	–	4.59	0.48	6.99	6.03	00.48	22.9	2.85	0.67	0.29	–
H	9.61	6.35	3.26	7.31	6.68	NA	0.32	8.89	7.78	0.56	34.2	4.75	1.04	0.36	2125
H3	9.61	6.11	3.5	7.31	–	6.22	0.55	8.89	7.78	0.56	34.2	4.75	1.04	0.36	–
P	12.26	8.5	3.76	9.53	8.87	NA	0.33	11.4	10.16	0.64	50.6	8.1	1.55	0.41	1430
P3	12.26	8.31	3.95	9.53	–	8.41	0.56	11.4	10.16	0.64	50.6	8.1	1.55	0.41	–

System	Hole size	Core D.	Hole-core	Inner tube				Rod properties					Rod-hole annulus		Depth
				OD	ID	ID with split tube	δ	OD	ID	δ	W	V	Volume capacity	δ	DE880
	mm			mm				mm			kg/(3 m)	l/m	l/m	mm	m
A	–	–	–	–	–	–	–	–	–	–	–	–	–	–	–
B	59.9	36.4	23.5	42.9	38	N/A	2.5	55.6	46	4.8	17.9	1.66	0.4	2.1	4060
N	75.7	47.6	28.1	55.6	49.9	N/A	2.9	69.9	60.3	4.8	22.9	2.85	0.67	2.9	3170
N2	75.7	50.7	25	57.2	52.4	N/A	2.4	69.9	60.3	4.8	22.9	2.85	0.67	2.9	–
N3	75.7	45.1	30.6	55.6	–	45.94	4.8	69.9	60.3	4.8	22.9	2.85	0.67	2.9	–
H	96.1	63.5	32.6	73.1	66.8	N/A	3.2	88.9	77.8	5.6	34.2	4.75	1.04	3.6	2125
H3	96.1	61.1	35	73.1	–	62.2	5.5	88.9	77.8	5.6	34.2	4.75	1.04	3.6	–

(continued)

Table 15.11 (continued)

System	Hole size	Core D.	Hole-core	Inner tube			δ	Rod properties					Rod-hole annulus		Depth
				OD	ID	ID with split tube		OD	ID	δ	W	V	Volume capacity	δ	DE880
	mm			mm				mm			kg/(3 m)	l/m	l/m	mm	m
P	122.6	85	37.6	95.3	88.7	N/A	3.3	114.3	101.6	6.4	50.6	8.1	1.55	4.1	1430
P3	122.6	83.1	39.5	95.3	–	84.1	5.6	114.3	101.6	6.4	50.6	8.1	1.55	4.1	–

aThe depth capacity with the strongest Sandvik drilling machine

Table 15.12 Specifications of thin-kerf wireline continuous coring systems [field units] (published courtesy of Sandvik 2013)

System	Hole size	Core D.	Hole-core	Inner tube			Rod (outer tube/barrel) properties			Rod-hole annulus		Depth capacity
	in.			OD	ID	δ	OD	ID	δ	Volume	δ	DE 880
				in.			in.			US gal/100 ft	in.	ft
TK56	2.23	1.53	0.7	1.79	1.59	0.10	2.1	1.73	0.19	2.5	0.06	18,060
TK66	2.64	1.96	0.68	2.25	2.06	0.10	2.52	2.32	0.1	2.7	0.06	15,480
TK66-3	2.64	1.81	0.83	2.25	2.06	0.10	2.52	2.32	0.1	2.7	0.06	15,480
TK76	3	2.24	0.76	2.43	2.29	0.07	2.88	2.48	0.2	2.9	0.06	10,690
TK76-3	3	2	1	2.25	2.06	0.10	2.88	2.48	0.2	2.9	0.06	10,690

Table 15.13 Specifications of thin-kerf wireline continuous coring systems [other units] (published courtesy of Sandvik 2013)

System	Hole Size	Core D.	Hole-core	Inner tube			Rod properties (outer tube/barrel) properties			Rod-hole annulus		Depth capacity
				OD	ID	δ	OD	ID	δ	Volume	δ	DE 880
	cm			cm			cm			l/m	cm	m
TK56	5.68	3.9	1.78	4.55	4.03	0.26	5.32	4.4	0.46	0.31	0.18	5500
TK66	6.71	5	1.71	5.71	5.23	0.24	6.4	5.89	0.26	0.33	0.15	4720
TK66-3	6.71	4.6	2.11	5.71	5.23	0.24	6.4	5.89	0.26	0.33	0.15	4720
TK76	7.63	5.7	1.93	6.17	5.82	0.175	7.32	6.3	0.51	0.36	0.15	3300
TK76-3	7.63	5.1	2.53	5.71	5.23	0.24	7.32	6.3	0.51	0.36	0.15	3300
System	Hole size	Core D.	Hole-core	Inner tube			Rod properties (outer tube/barrel) properties			Rod-hole annulus		Depth capacity
				OD	ID	δ	OD	ID	δ	Volume	δ	DE 880
	mm			mm			mm			l/m	mm	m
TK56	56.8	39	17.8	45.5	40.3	2.6	53.2	44	4.6	0.31	1.8	5500
TK66	67.1	5	17.1	57.1	52.3	2.4	64	58.9	2.6	0.33	1.55	4720
TK66-3	67.1	46	21.1	57.1	52.3	2.4	64	58.9	2.6	0.33	1.55	4720
TK76	76.3	57	19.3	61.7	58.2	1.75	73.2	63	5.1	0.36	1.55	3300
TK76-3	76.3	51	25.3	57.1	52.3	2.4	73.2	63	5.1	0.36	1.55	3300

Table 15.14 Wireline continuous coring systems [field units] (published courtesy of Boart Longyear)

system type	Hole size	Core D.	Inner tube			Rod (outer tube/barrel) properties					Rod/hole annulus		Depth capacity
			OD	ID	δ	OD	ID	δ	Weight	Volume or capacity	Volume capacity	δ	
	in.	in.	in.			in.			lb/10 ft	US gal/100 ft	US gal/100 ft	in.	ft
ARQ™TK/U	1.89	1.2	1.66	1.46	0.09	1.76	1.47	0.15	2.53	9	1.88	0.06	4900
BQ™/U	2.36	1.43	2.1	1.9	0.09	2.19	1.81	0.19	4.2	13	3.01	0.09	4900
BRQ™a/U[b]	2.36	1.43	2.1	1.9	0.09	2.19	1.81	0.19	4.2	13	3.01	0.09	9800
BRQ™TK/U	2.36	1.6	2.1	1.9	0.09	2.2	1.9	0.15	3.3	15	2.93	0.08	4900
NQ™/U	2.98	1.88	2.65	2.45	0.09	2.75	2.38	0.19	5.23	23	5.34	0.12	4900
NQ3™c	2.98	1.77	2.65	2.45	0.09	2.75	2.38	0.19	5.23	23	5.34	0.12	4900
NQ™TK[d]/U	2.98	2	2.65	2.45	0.09	2.75	2.38	0.19	5.23	23	5.34	0.12	4900
NQ™ V-Wall™e	2.98	1.88	2.65	2.45	0.09	2.75	2.44	0.16	4.5	24	5.34	0.12	6500
NRQ™/U	2.98	1.88	2.65	2.45	0.09	2.75	2.37	0.19	5.23	23	5.34	0.12	9800
NRQ™ V-Wall™	2.98	1.88	2.65	2.45	0.09	2.75	2.44	0.16	4.5	24	5.34	0.12	10,800
HQ™/U	3.78	2.5	3.4	3.16	0.11	3.5	3.06	0.22	7.7	38	8.3	0.14	4900
HQ™3	3.78	2.41	3.4	3.16	0.11	3.5	3.06	0.22	7.7	38	8.3	0.14	4900
HQ™ V-Wall™	3.78	3.19	3.4	3.16	0.11	3.5	3.19	0.16	6	40	8.3	0.14	4900
HRQ™	3.78	2.41	3.4	3.16	0.11	3.5	3.07	0.22	7.7	40	8.3	0.14	8200
HRQ™ V-Wall™	3.78	2.41	3.4	3.16	0.11	3.5	3.19	0.16	6	40	8.3	0.14	10,000

(continued)

Table 15.14 (continued)

system type	Hole size in.	Core D.	Inner tube OD in.	ID	δ	Rod (outer tube/barrel) properties OD in.	ID	δ	Weight lb/10 ft	Volume or capacity US gal/100 ft	Rod/hole annulus Volume capacity US gal/100 ft	δ in.	Depth capacity ft
PHD	4.83	3.35	4.37	4.05	0.15	4.5	4	0.25	11.7	65	12.87	0.17	4900
PQTM3	4.83	3.27	4.37	4.05	0.15	4.5	4	0.25	11.7	65	12.87	0.17	4900
PHD V-Wall™	4.83	3.35	4.37	4.05	0.15	4.5	4.19	0.16	8.21	70	12.87	0.17	6500

Q, RQ, and V-Wall are registered trademarks of Boart Longyear and are used with permission

[a]With a different "reverse" thread design (stronger) they are made stronger, but with the same rod dimensions as Q and the same core size

[b]Underground

[c]Triple tube system. In triple tube systems, the core diameter is smaller

[d]In thin kerf systems, the core diameter is higher. NQTK system is used with same NQ drill rods

[e]V-Wall rods are internally upset rods. These rods maintain the specified outside diameter over their entire length and standard wall thickness at each threaded end. The inside diameter gradually increases over the length of the rod, achieving a thinner wall thickness at mid-body

Table 15.15 Wireline continuous coring systems [other units] (published courtesy of Boart Longyear)

System type	Hole size	Core D.	Inner tube			Rod (outer tube/barrel) properties					Rod/hole annulus		Depth capacity
			OD	ID	δ	OD	ID	δ	Weight	Volume or capacity	Volume capacity	δ	
	cm		cm			cm			kg/3 m	l/m	l/m	cm	m
ARQ™TK/U	4.8	3.05	4.23	3.73	¼	4.48	3.74	0.37	11.3	1.09	0.23	0.16	1500
BQ™/U	6	3.64	5.34	4.84	¼	5.59	4.61	0.49	18.75	1.67	0.37	0.21	1500
BRQ™a/Ub	6	3.64	5.34	4.84	¼	5.59	4.61	0.49	18.75	1.67	0.37	0.21	3000
BRQ™TK/U	6	4.07	5.35	4.84	¼	5.6	4.83	0.38	14.85	1.83	0.36	0.2	1500
NQ™/U	7.57	4.76	6.74	6.24	¼	6.99	6.2	0.39	23.37	2.86	0.66	0.29	1500
NQ™ 3c	7.57	4.5	6.74	6.24	¼	6.99	6.2	0.39	23.37	2.86	0.66	0.29	1500
NQ™TK/U	7.57	5.06	6.74	6.24	¼	6.99	6.2	0.39	23.37	2.86	0.66	0.29	1500
NQ™ V-Wall™d	7.57	4.76	6.74	6.24	¼	6.99	6.2	0.39	20.42	2.97	0.66	0.29	2000
NRQ™/U	7.57	4.76	6.74	6.24	¼	6.99	6.02	0.48	23.37	2.86	0.66	0.29	3000
NRQ™ V-Wall™	7.57	4.76	6.74	6.24	¼	6.99	6.2	0.39	20.42	2.97	0.66	0.29	3300
HQ™/U	9.6	6.35	8.64	8.04	0.3	8.89	7.78	0.55	34.3	4.8	1.03	0.36	1500
HQ™3	9.6	6.11	8.64	8.04	0.3	8.89	7.78	0.55	34.3	4.8	1.03	0.36	1500
HQ™ V-Wall™	9.6	6.35	8.64	8.04	0.3	8.89	8.1	0.39	27.3	5.14	1.03	0.36	1500
HRQ™	9.6	6.35	8.64	8.04	0.3	8.89	7.79	0.55	34.3	4.8	1.03	0.36	2500
HRQ™ V-Wall™	9.6	6.35	8.64	8.04	0.3	8.89	8.1	0.39	27.3	5.14	1.03	0.36	3050
PHD	12.26	8.5	11.1	10.3	0.4	11.4	10.1	0.65	52.2	8.14	1.6	0.43	1500
PQ™3	12.26	8.3	11.1	10.3	0.4	11.4	10.1	0.65	52.2	8.14	1.6	0.43	1500
PHD V-Wall™	12.26	8.5	11.1	10.3	0.4	11.4	10.6	0.4	37.3	8.8	1.6	0.43	2000

Q, RQ, and V-Wall are registered trademarks of Boart Longyear and are used with permission

aWith reverse threads (stronger)
bUnderground
cTriple tube
dWhich has the highest depth capacity

Table 15.16 Wireline continuous coring systems [other units, cont.] (published courtesy of Boart Longyear)

System type	Hole size	Core D.	Inner tube			Rod (outer tube/barrel) properties					Rod/hole annulus		Depth capacity
			OD	ID	δ	OD	ID	δ	Weight	Volume or capacity	Volume capacity	δ	
	mm	mm	mm			mm			kg/3 m	l/m	l/m	mm	m
ARQ™TK/U	48	30.5	42.3	37.3	2.5	44.8	37.4	3.7	11.3	1.09	0.23	1.6	1500
BQ™/U	60	36.4	53.4	48.4	2.5	55.9	46.1	4.9	18.75	1.67	0.37	2.1	1500
BRQ™/U[b]	60	36.4	53.4	48.4	2.5	55.9	46.1	4.9	18.75	1.67	0.37	2.1	3000
BRQ™TK/U	60	40.7	53.5	48.4	2.5	56	48.3	3.8	14.85	1.83	0.36	2	1500
NQ™/U	75.7	47.6	67.4	62.4	2.5	69.9	62	3.9	23.37	2.86	0.66	2.9	1500
NQ™3[c]	75.7	45	67.4	62.4	2.5	69.9	62	3.9	23.37	2.86	0.66	2.9	1500
NQ™TK/U	75.7	50.6	67.4	62.4	2.5	69.9	62	3.9	23.37	2.86	0.66	2.9	1500
NQ™ V-Wall™[d]	75.7	47.6	67.4	62.4	2.5	69.9	62	3.9	20.42	2.97	0.66	2.9	2000
NRQ™/U	75.7	47.6	67.4	62.4	2.5	69.9	60.2	4.8	23.37	2.86	0.66	2.9	3000
NRQ™ V-Wall™	75.7	47.6	67.4	62.4	2.5	69.9	62	3.9	20.42	2.97	0.66	2.9	3300
HQ™/U	96	63.5	86.4	80.4	3	88.9	77.8	5.5	34.3	4.8	1.03	3.6	1500
HQ™3	96	61.1	86.4	80.4	3	88.9	77.8	5.5	34.3	4.8	1.03	3.6	1500
HQ™ V-Wall™	96	63.5	86.4	80.4	3	88.9	81	3.9	27.3	5.14	1.03	3.6	1500
HRQ™	96	63.5	86.4	80.4	3	88.9	77.9	5.5	34.3	4.8	1.03	3.6	2500
HRQ™ V-Wall™	96	63.5	86.4	80.4	3	88.9	81	3.9	27.3	5.14	1.03	3.6	3050
PHD	122.6	85	111	103	4	114	101	6.5	52.2	8.14	1.6	4.3	1500
PQ™3	122.6	83	111	103	4	114	101	6.5	52.2	8.14	1.6	4.3	1500
PHD V-Wall™	122.6	85	111	103	4	114	106	4	37.3	8.8	1.6	4.3	2000

Q, RQ, and V-Wall are registered trademarks of Boart Longyear and are used with permission

[a]With reverse threads (which are stronger than normal threads)

[b]Underground

[c]Triple tube system

[d]Has the highest depth capacity

Table 15.17 Patent information corresponding to some important coring systems

Subject	More info	Patent topic	US patent no	publication date	Inventors	Applicant
1. Coring (General)		Combination drill and core bit	US 2,708,103 A	May 10, 1955	Williams Jr Edward B	Williams Jr Edward B
		Core bit	US3,032,130 A	May 1, 1962	Elzey Lloyd J	Elzey Lloyd J
		Apparatus for taking core samples	US 4,981,183 A	Jan 1, 1991	Gordon A. Tibbitis	Baker Hughes
2. Wireline Continuous		Combination drill and core bit	US 2,708,103 A	May 10, 1955	Williams Jr Edward B	Williams Jr Edward B
		Wireline core barrel	US 3,127,943 A	Apr 7, 1964	Takeshi Mori	Christensen Diamond Prod Co
	Wireline plus reverse circulation	Coring apparatus with hydraulically retrievable inner core barrel	US 3,481,412 A	Dec 2, 1969	Rowley David S	Christensen Diamond Prod Co
		Drilling apparatus, particularly wireline core drilling apparatus	US 5,339,915 A	Aug 23, 1994	Irwin J. Laporte, Amos J. Watkins	Jks Boyles International, Inc.
	Inner Tube Latching	Wireline core drilling apparatus	US 5,020,612 A	Jun 4, 1991	David Stanley Williams	Boart International Limited
	Hydraulic latching of inner tube	Coring assembly and method	US 5,351,765 A	Oct 4, 1994	Ronald D. Ormsby	Baroid Technology, Inc.
		Bit-stabilized combination coring and drilling system	US 5,568,838 A	Oct 29, 1996	Barry W. Struthers, Pierre E. Collee	Baker Hughes GE
		Apparatus and method for coring and/or drilling	US 6,712,158 B2	30 Mar 2004	Terence Alexander Moore	Corpro

(continued)

Table 15.17 (continued)

Subject	More info	Patent topic	US patent no	publication date	Inventors	Applicant
	Specialty drill pipes	Drilling system and method suitable for coring and other purposes	6,736,224	May 18, 2004	Douglas Kinsella	Corpro
		Apparatus and methods for continuous coring	US 8,162,080 B2	24 Apr 2012	Homero Castillo	Baker Hughes GE
	Core Size Enhanced	Drilling system for enhanced coring and method	US 8,579,049 B2	Nov 12, 2013	Douglas Kinsella	Corpro
3. Invasion-Mitigation Coring		Low-invasion coring fluid	US 3,314,489 A	18 Apr 1967	Raymond A Humphrey	Exxon Production Research Co
		Method and apparatus for pressure coring with non-invading gel	US 5,482,123 A	Jan. 9, 1996	Pierre E. Collee	Baker Hughes GE
4. Oriented Coring		Core orienting apparatus and method	US 2,657,013 A	Oct. 27, 1953	Edward F. Brady	Eastman Oil Well Survey Co.
		Method and apparatus for orienting cores	US 3,059,707 A	Oct. 23, 1962	Thomas M. Frisby	Eastman Oil Well Survey Co.
		Orientation coring tool	US 3363703 A	Jan 16, 1968	Parkes Shewmake	Parkes Shewmake
		Orientation coring tool	US 3,363,703 A	Jan 16, 1968	Parkes Shewmake	Parkes Shewmake
		Core barrel for obtaining oriented cores	EP 0253,473 A2	Jan. 20, 1988	Terence Alexander Moore	Diamant Boart Stratabit Limited (applicant)
5. Pressure Coring	Conventional	Pressure core barrel	US 3,548,958 A	Dec 22, 1970	Blackwell Robert J, Rumble Robert C	Exxon Production Research Co

(continued)

Table 15.17 (continued)

Subject	More info	Patent topic	US patent no	publication date	Inventors	Applicant
	Combined with gel coring	Method and apparatus for pressure coring with non-invading gel	US 5,482,123 A	Jan 9, 1996	Pierre E. Collee	Baker Hughes GE
	Wireline	Apparatus for recovering core samples under pressure	US 6,230,825 B1 US 6,378,631 B1	-May 15, 2001 -April 30, 2002	James T. Aumann, Craig R. Hyland	James T. Aumann, Craig R. Hyland
	Wireline	Device and method for extracting a sample while maintaining a pressure that is present at the sample extraction location	WO 2,013,060,720 A3	Jun 20, 2013	Tobias Rothenwänder et al.	Corsyde, Technische Universität Berlin
	Wireline (pseudo-pressure)	Pressure Coring Assembly and Method	EP2,686,515 A2	Jan. 22, 2014	Douglas Kinsella	Corpro
	Wireline (pseudo-pressure)	Tight gas formation pressure determination method	WO 2,015,108,880 A1	Jul 23, 2015	Donald Westacott, Luis F. Quintero	Halliburton
6. LWC		Means and method for facilitating measurements while coring	US 4,601,354 A	July 22, 1986	Frank L. Campbell, Dean C. Barnum, William C. Corea	Chevron Research Co.
		Method and apparatus for simultaneous coring and formation evaluation	US 6,006,844 A	Dec 28, 1999	Luc Van Puymbroeck, John W. Harrel, Michael H. Johnson, Pierre E. Collee	Baker Hughes GE

(continued)

Table 15.17 (continued)

Subject	More info	Patent topic	US patent no	publication date	Inventors	Applicant
		Downhole in situ measurement of physical and or chemical properties including fluid saturations of cores while coring	US 6003620 A	Dec 21, 1999	Mukul M. Sharma, Roger T. Bonnecaze, Bernhard Zemel	Advanced Coring Technology
		Logging-while-coring method and apparatus	US 2005/0199393 A1	Sep. 15, 2005	David S. Goldberg, Gregory J. Myers	The Trustees of Columbia University, New York, NY
	(Real-time coring, core removal possibility)	Intelligent coring system	WO 2014012781 A2	Jan 23, 2014	Per-Erik Berger	Coreall As
		Coring apparatus with sensors	US 8,146,684 B2	Feb. 22, 2012	Phillipe Cravatte	Corpro
	(Strain and compression of inner tube, etc.)	Monitoring apparatus for core barrel operations	US 7,878,269 B2	Feb 1, 2011	Phillipe Cravatte	Corpro
		Apparatus and methods for monitoring a core during coring operations	US 8,797,035 B2	Aug 5, 2014	Michael S. Bittar, Gary E. Weaver	Halliburton Energy Services
7. Sponge Coring		Apparatus and methods for sponge coring	US 6,719,070 B1	April. 13, 2004	Luc Van Puymbroeck, Bob T. Wilson, Holger Stibbe, Hallvard S. Hatloy	Baker Hughes GE
		Apparatus and methods for sponge coring	US 7,004,265 B2	2006	Luc Van Puymbroeck, Bob T. Wilson, Holger Stibbe, Hallvard S. Hatloy	Baker Hughes GE
	(Fiber in lieu of sponge)	Methods and apparatus for coring	WO 2,013,052,165 A2	11 Apr 2013	Bobby Talma Wilson	NOV

(continued)

Table 15.17 (continued)

Subject	More info	Patent topic	US patent no	publication date	Inventors	Applicant
8. Motor Coring (with turbine)		Drill tool comprising a core barrel and a removable central portion	US 3,951,219 A	Apr 20, 1976	Abel C. Cortes	Compagnie Francaise Des Petroles
		Method for directional coring	US 5,029,653 A	Jul 9, 1991	Rainer Jurgens, Johann van Es	Baker Hughes GE
	(Electrical Motor and piston)	Coring assembly for mounting on the end of a drill string	US 5,103,921 A	Apr 14, 1992	Robert L. Zeer, Alex Mihai	Sidetrack Coring Systems Inc
		Downhole coring device	EP 1,334,260 B1		Peter Nicolaas Looijen, Herman Maria Zuidberg	Fugro Engineers B.V.
9. Under-Balanced Coring		Underbalanced coring tool	CN 201,581,840 U	Sep 15, 2010	–	–
10. Coiled-Tubing Coring		System and method for conducting drilling and coring operations	US 20,130,056,276 A1	Mar 7, 2013	Denis Rousseau, Jeremy Myers, Richard Havinga	Denis Rousseau, Jeremy Myers, Richard Havinga

Eastman Company (which was active in directional drilling and MWD) was merged to INTEQ and is now part of Baker Hughes GE (BHI)

Appendix

The input data for modelling the safe tripping rates shown in Fig. 8.3 (Ashena et al. 2018)

Parameter	Value	Evaluation method
Dimensions		
Initial bottom-hole depth [m]	2000	
Hole size [in.]	3½	
Diameter of core [in.]	2	
Diameter of core barrel [in.]	2¾	
Number of inner barrel joints	2	
Length of each joint [ft]	20	
Length of the core [ft]	$2 \times 20 = 40$	
Rock properties		
Porosity, ϕ [%]	30	Estimated/measurable
Permeability of core, K [mD]	10^{-4} to 1	Estimated/measurable
Gas properties		
Viscosity of gas, μ_g [cp]	0.04 (surface)	Estimated/measurable
Molecular weight of gas, M_g	16 (methane)	Depending on the gas
Specific gravity of gas (surface)	0.67	Depending on the gas
Water properties		
Specific gravity of water	1	Estimated/measurable
Viscosity of water, μ_w [cp]	1	Estimated/measurable
Compressibility		
Compressibility of rock, C_r [1/Pa]	5×10^{-10}	Estimated/measurable
Compressibility of gas (surface), C_g [1/Pa]	7.2×10^{-8}	Estimated/measurable
Compressibility of water, C_w [1/Pa]	5×10^{-10}	Estimated/measurable
Interstitial water saturation, $S_{w,i}$	20%	Estimated/measurable
Total compressibility, $C_{t,g}$ [1/Pa] (gas-bearing core)	7.9×10^{-6}	$C_t = C_r + S_w C_w + S_g C_g$ (Ahmed and McKinney 2005)

(continued)

© Springer International Publishing AG, part of Springer Nature 2018
R. Ashena and G. Thonhauser, *Coring Methods and Systems*,
https://doi.org/10.1007/978-3-319-77733-7

(continued)

Parameter	Value	Evaluation method
Total compressibility, $C_{t,w}$ [1/Pa] (water-bearing core)	10^{-9}	$C_t = C_r + S_w C_w + S_g C_g$ (Ahmed and McKinney 2005)
Hydraulic diffusivity		
Hydraulic diffusivity, η [m^2/s] (gas-bearing at surface)	10^{-5} to 10^{-9} (depending on K)	$\eta = 9.869 \times 10^{-13} \frac{K}{\varphi \mu_g C_{t,g}}$ (Ahmed and McKinney 2005)
Hydraulic diffusivity, η [m^2/s] (water-bearing at surface)	6×10^{-3} to 6×10^{-7} (depending on K)	$\eta = 9.869 \times 10^{-13} \frac{K}{\varphi \mu_w C_{t,w}}$ (Ahmed and McKinney 2005)
Hydraulic diffusivity, η [m^2/s] (oil-bearing at surface)		
Oil-related properties		
Oil isothermal compressibility, C_o [1/MPa]	2.55×10^{-3}	
Oil bubble point pressure, P_{ob} [MPa]	30	
Oil formation volume factor at bubble point, $B_{o,b}$	1.5	
Oil viscosity at bubble point, $\mu_{g,b}$ [Pa s]	3.5×10^{-4}	
Gas critical pressure, $P_{g,cr}$ [MPa]	22	
Henry's solubility coefficient, h	0.7	
Fluid compressibility factor, J	0.41	
Thermal properties		
Thermal expansion coefficient, α_m [1/°C]	1.2×10^{-5}	Estimated/measurable (Timoshenko 1934)
Thermal diffusivity coefficient, η_T [m^2/s]	8×10^{-7}	Estimated
Geothermal gradient [°C/m]	0.044	Estimated/measurable
Thermal coefficient, $C_{Thermal}$ [MPa/(m/s)]	0.69	Calculated
Mechanical properties		
Uniaxial compressive strength, UCS [Mpa]	20	Measurable/estimated
Tensile strength, T.S. [Mpa]	2	T.S. $=$ UCS$/m$, $m = 7 - 15$ (Jaeger et al. 2007)
Biot's coefficient, α	0.7	0.6–0.7 (for shales)
Poisson's ratio, v	0.3	Estimated/measurable
Bulk modulus, K [GPa]	5	$K = \frac{E}{3(1-2v)}$ (Wang 2000; Zoback 2010)
Mud properties		
Mud weight, MW [kg/m^3]	1078	MW[ppg] $\times$ 119.826 $=$ MW $\left[\frac{kg}{m^3}\right]$
Mud cake thickness, t_{mc} [mm]	2	Estimated/measurable

(continued)

(continued)

Parameter	Value	Evaluation method
Yield point, YP [Ibf/100 ft^2]		12
Plastic viscosity, PV [cp]		30
R_{total}/R_{core}	1.08	Estimated/measurable
$K_{mud-cake}/K_{core}$	0.8	Estimated/measurable
Mud cake coefficient, C_{mc} (%)		Calculated
Swab coefficients		
C_{Swab}	0.0537	Calculated
K_{Swab}	7.6039	Calculated